AF770456

Practice Problems

to accompany

Physics

FOR SCIENTISTS AND ENGINEERS WITH MODERN PHYSICS

Fourth Edition

Raymond Serway

James Madison University

Saunders College Publishing

Harcourt Brace College Publishers

Fort Worth Philadelphia San Diego New York Orlando Austin San Antonio Toronto Montreal London Sydney Tokyo

Printed in the United States of America.

Serway & Van Wyk; Practice Problems to accompany Physics for Scientists and Engineers. Serway

ISBN 0-03-018238-7

567 017 987654321

Preface

This collection of 833 problems is taken from the third edition of *Physics for Scientists and Engineers with Modern Physics* by Raymond A. Serway. This ancillary serves as a supplement to the fourth edition of this textbook and can be used as a source of additional practice in preparing for examinations. Most of the problems contained in this supplement are at the first level, and answers to all problems are provided at the back of the booklet.

Acknowledgements

I thank Steve Van Wyk for providing the answers to the problems in this supplement and Jennifer Serway for her assitance in preparing the manuscript.

Raymond A. Serway
James Madison University
Harrisonburg, VA 22807

August, 1995

Table of Contents

Table of Contents

Chapter 1

1. A solid sphere is to be made out of copper, which has a density of 8.93 g/cm^3. If the mass of the sphere is to be 475 g, what radius must it have?

2. Assume that it takes 7.0 minutes to fill a 30-gal gasoline tank. (a) Calculate the rate at which the tank is filled in units of gallons per second. (b) Calculate the rate at which the tank is filled in units of cubic meters per second. (c) Determine the time in hours required to fill a one-cubic-meter volume at the same rate.

3. Show that the equation $v^2 = v_0^2 + 2ax$ is dimensionally correct, where v and v_0 represent velocities, a is acceleration, and x is a distance.

4. Suppose that the displacement of a particle is related to time according to the expression $s = ct^3$. What are the dimensions of the constant c ?

5. An object in the shape of a rectangular parallelepiped measures 2.0 in. × 3.5 in. × 6.5 in. Determine the volume of the object in m^3.

6. Estimate the age of the Earth in years using the data in Table 1.3 and the appropriate conversion factors.

7. Using the fact that the speed of light in free space is about 3.00×10^8 m/s, determine how many miles a pulse from a laser will travel in one hour.

8. A painter is to cover the walls in a room 8.00 ft high and 12.0 ft along each side. What surface area in square meters must she cover?

9. The distance between our Sun and the star nearest it is about 4.0×10^{13} km. If the Sun (diameter = 1.4×10^9 m) and this nearest star are each represented by a cherry pit 7.0 mm in diameter, how far apart should the two pits be placed to represent the Sun and its neighbor to scale?

10. Soft drinks are commonly sold in aluminum containers. Estimate the number of such containers used each year by U.S. consumers. Approximately how many tons of aluminum does this represent?

11. The United States consumes petroleum at a rate of about 6×10^9 barrels per year. Assuming a barrel has a length of 1 m, compare the length of 6 billion barrels, laid end-to-end, with the coast-to-coast distance of the United States (about 4000 km).

12. Calculate (a) the circumference of a circle of radius 3.5 cm and (b) the area of a circle of radius 4.65 cm.

Chapter 2

1. An athlete swims the length of a 50-m pool in 20 s and makes the return trip to the starting position in 22 s. Determine his average velocity in (a) the first half of the swim, (b) the second half of the swim, and (c) the round trip.

2. Using Figure P2.2. determine (a) the average velocity between $t = 2.0$ s and $t = 5.0$ s, and (b) the instantaneous velocity at $t = 3.0$ s.

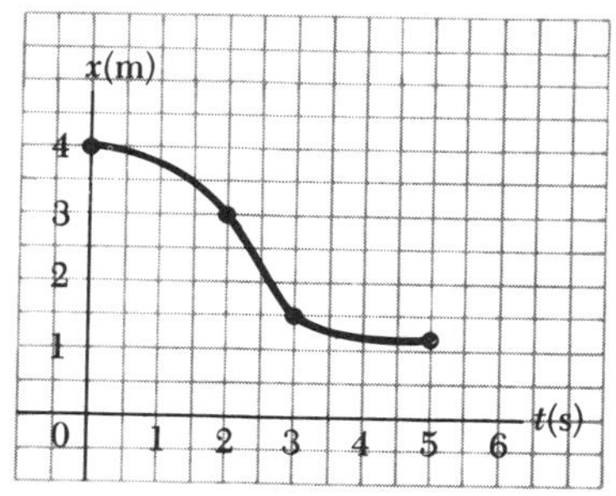

Figure P2.2

3. A car traveling in a straight line has a velocity of 30 m/s at some instant. Two seconds later its velocity is 25 m/s. What is its average acceleration in this time interval?

4. The velocity-time graph for an object moving along the x axis is shown in Figure P2.4. (a) Plot a graph of the acceleration versus time. (b) Determine the average acceleration of the object in the time intervals $t = 5.0$ s to $t = 15.0$ s and $t = 0$ to $t = 20.0$ s.

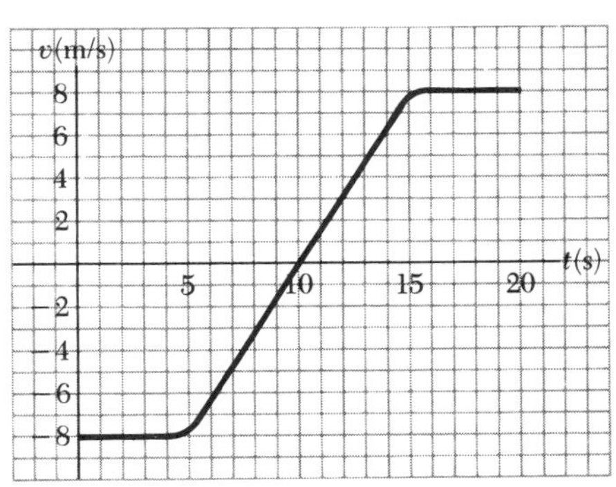

Figure P2.4

Chapter 2

5. When struck by a club, a golf ball initially at rest acquires a speed of 31.0 m/s. If the ball is in contact with the club for 1.17 ms, what is the magnitude of the average acceleration of the ball?

6. An automobile traveling initially at a speed of 60.0 m/s is accelerated uniformly to a speed of 85.0 m/s in 12.0 s. How far does the automobile travel during the 12.0-s interval?

7. The initial velocity of an electron is -2.0×10^5 m/s. If its acceleration is 4.2×10^{14} m/s^2, (a) how far does it travel before momentarily coming to rest? (b) How long will it take to undergo a displacement of 5.0 cm?

8. A locomotive slows from 26.0 m/s to zero in 18.0 s. What distance does it travel?

9. A body is traveling at 16.0 m/s and comes to rest after undergoing a uniform negative acceleration for 40.0 m. (a) What is the acceleration of this body? (b) How long does it take to stop?

10. A go-cart travels the first half of a 100-m track with a constant speed of 5.00 m/s. In the second half of the track, it experiences a mechanical problem and slows down at 0.200 m/s^2. How long does it take the go-cart to travel the 100-m distance?

11. A railroad car is released from a locomotive on an incline. When the car reaches the bottom of the incline, it has a speed of 30 mi/h, at which point it passes through a retarder track that slows it down. If the retarder track is 30 ft long, what negative acceleration must it produce to stop the car?

Chapter 2

12. A stone is thrown upward from the edge of a cliff 18.0 m high. It just misses the cliff on the way down and hits the ground below with a speed of 18.8 m/s. (a) With what velocity was it released? (b) What is its maximum distance from the ground during its flight?

13. The velocity of an object is given by $v = at + b$, where $a = 2.40 \text{ m/s}^2$ and $b = 30.0 \text{ m/s}$. Plot this velocity versus time on graph paper from $t = 0$ to $t = 10.0$ s, and calculate the distance covered during that time interval by determining the area under the curve.

14. A particle is moving along the x axis. Its velocity as a function of time is given by $v = 5.00 + 10.0t$, where v is in meters per second. The position of the particle at $t = 0$ is 20.0 m. Find (a) the particle's acceleration as a function of time, (b) its position as a function of time, and (c) its velocity at $t = 0$.

15. Given the position function $x(t) = \frac{1}{4}at^2 + 2bt$ where $a = 1.00 \text{ m/s}^2$ and $b = 1.00 \text{ m/s}$, evaluate the average velocity $\overline{v}$ during elapsed time intervals starting at $t = 2.00$ s for which $\Delta t = 1.00$ s, 0.500 s, 0.100 s, 0.0100 s, and 0.00100 s.

16. The velocity of a rocket during the launch stage is given by $v = bt - kt^2$, where $b = 30.0 \text{ m/s}^2$ and $k = 0.500 \text{ m/s}^3$. Determine the average acceleration of the rocket from (a) $t_i = 0$ s to $t_f = 1.00$ s and (b) $t_i = 5.00$ s to $t_f = 6.00$ s.

17. By matching geological formations on either side of the San Andreas fault in California, geologists find a net displacement of 325 km occurring over a period of about 20 million years. Find the average rate of displacement in centimeters per year. (Note: In 1987 the rate of slippage near the Hollister area was 6.0 cm/y, about the rate of growth of your fingernails.)

Chapter 2

18. A ball is released from rest from a height of 2.00 m, and the time it takes to reach the floor is measured with a stopwatch. The height of the ball is measured to an accuracy of 2 cm and the experiment is repeated 25 times to give an average time of fall of 0.63 s. What value of g do you obtain from these data? Compare your value with the accepted value of 9.80 m/s^2.

19. A particle undergoes a varying acceleration. The velocity is measured at 0.5-s intervals and is tabulated below. (a) Determine the average acceleration in each interval. (b) Use a numerical integration procedure to determine the position of the particle at the end of each time interval. Assume the initial position of the particle is zero.

t (s)	0	0.5	1.0	1.5	2.0	2.5	3.0	3.5	4.0	4.5	5.0
v (m/s)	0	1.0	3.0	4.5	7.0	9.5	10.5	12.0	14.0	15.0	17.5

20. The acceleration of a particle moving along the x axis varies with position according to the expression $a = a_0 e^{-bx}$, where $a_0 = 3.00$ m/s^2 and $b = 1.00\ m^{-1}$. For a particle starting at rest from the origin, use a numerical integration method to find the position of the particle at $t = 2.37$ s. The accuracy of your calculation should be at least 1%.

21. A particle moving along the x axis undergoes an acceleration $a = \sqrt{3 + t^3}$ m/s^2. Use a numerical integration method to find the position and velocity of the particle at $t = 5.70$ s to within 1% accuracy.

Chapter 3

1. A point in the xy plane has cartesian coordinates (-3.00, 5.00) m. What are the polar coordinates of this point?

2. A shopper pushing a cart through a store begins at point A. He moves 40.0 m down one aisle, then makes a 90.0° turn and moves 15.0 m. He then makes another 90.0° turn and moves 20.0 m, ending up at point B. What is the magnitude and direction of the resultant displacement from A to B? The directions moved in the 90.0° turns are not given. As a result, could there be more than one answer?

3. Vector **A** is 3.00 units in length and points along the positive x axis. Vector **B** is 4.00 units in length and points along the negative y axis. Use graphical methods to find the magnitude and direction of the vectors (a) **A** + **B**, (b) **A** - **B**.

4. Walking along streets laid out in a square grid pattern, a child walks 2.0 blocks west, 3.0 blocks north, then 2.0 blocks west. (a) Find the total distance traveled. (b) Find the displacement (magnitude and direction) from the starting point.

5. A jogger runs 100 m due west, then changes direction for the second leg of the run. At the end of the run, she is 175 m away from the starting point at an angle of 15.0° north of west. What was the direction and length of her second displacement? Use graphical techniques.

6. Three vectors are given by **A** = **i** + 3**j**, **B** = 2**i** - **j**, and **C** = 3**i** + 5**j**. Find (a) the sum of the three vectors and (b) the magnitude and direction of the resultant vector.

Chapter 3

7. Given the displacement vectors **A** = (2.0**i** + 3.0**j**- 4.0**k**) m and **B** = (-2.0**i** - 2.0**j** + 2.0**k**) m, find the displacement vectors (a) **C** = **A** + **B** and (b) **D** = **A** - **B**, expressing each in terms of its Cartesian components and also its magnitude.

8. A vector **A** has a positive x component of 4 units and a negative y component of 2 units. What second vector **B** when added to **A** will produce a *resultant* vector three times the magnitude **A** of directed in the positive y direction?

9. Find the magnitude and direction of a displacement vector having x and y components of -5.00 m and 3.00 m, respectively.

10. A particle moves from a point in the xy plane having cartesian coordinates (- 3.00, - 5.00) m to a point with coordinates (- 1.00, 8.00) m. (a) Write vector expressions for the position vectors in unit-vector form for these two points. (b) What is the displacement vector?

Chapter 4

1. A particle initially located at the origin has an acceleration of $\mathbf{a} = 3.00\mathbf{j}\ \text{m/s}^2$ and an initial velocity of $\mathbf{v}_0 = 5.00\mathbf{i}\ \text{m/s}$. Find (a) the position vector and velocity at any time t and (b) the coordinates and speed of the particle at $t = 2.00$ s.

2. A student stands at the edge of a cliff and throws a stone horizontally over the edge with a speed of 18.0 m/s. The cliff is 50.0 m above a flat, horizontal beach, as shown in Figure P4.2. How long after being released does the stone strike the beach? With what speed and angle of impact does it land?

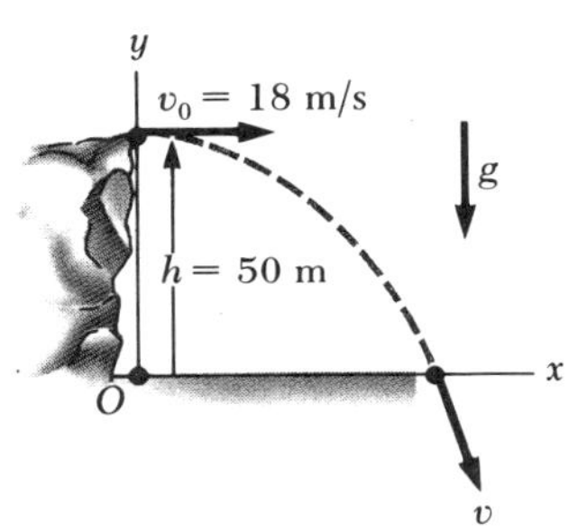

Figure P4.2

3. A golfer wants to drive a golf ball a distance of 310 yards (283 m). If the 4-wood launches the ball at 15.0° above the horizontal, what must be the initial speed of the ball to achieve the required distance?

4. During the 1968 Olympics in Mexico City, Bob Beamon executed a record long jump. The horizontal distance he achieved was 8.90 m. The center of his body started at an elevation of 1.00 m, reached a maximum height of 1.90 m, and finished at 0.150 m. From these data determine (a) his time of flight, (b) his horizontal and vertical velocity components at takeoff, and (c) his takeoff angle.

Chapter 4

5. It has been said that in his youth George Washington threw a silver dollar across a river. Assuming that the river was 75.0 m wide, (a) what *minimum initial* speed was necessary to get the coin across the river and (b) how long was the coin in flight?

6. The maximum horizontal distance a certain baseball player is able to hit the ball is 150 m. On one pitch, this player hits the ball in such a way that it has the same initial speed as his maximum-distance hit but makes an angle of 20.0° with the horizontal. Where will this ball strike the ground relative to home plate?

7. An astronaut on a strange planet finds that she can jump a maximum horizontal distance of 30.0 m if her initial speed is 9.00 m/s. What is the free-fall acceleration on the planet?

8. Find the acceleration of a particle moving with a constant speed of 8.00 m/s in a circle 2.00 m in radius.

9. A particle moves in a circular path 0.400 m in radius with constant speed. If the particle makes five revolutions in each second of its motion, find (a) the speed of the particle and (b) its acceleration.

10. An automobile whose speed is increasing at a rate of 0.600 m/s^2 travels along a circular road of radius r = 20.0 m. When the instantaneous speed of the automobile is 4.00 m/s, find (a) the tangential acceleration component, (b) the centripetal acceleration component, and (c) the magnitude and direction of the total acceleration.

Chapter 4

11. At some instant of time, a particle moving counterclockwise in a circle of radius 2.00 m has a speed of 8.00 m/s and its total acceleration is directed as shown in Figure 4.11. At this instant, determine (a) the centripetal acceleration of the particle, (b) the tangential acceleration, and (c) the magnitude of the total acceleration.

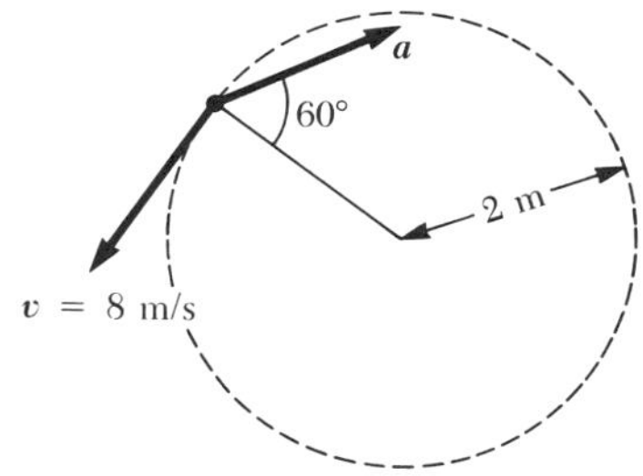

Figure 4.11

12. A car travels north with a speed of 60.0 km/h on a straight highway. A truck travels in the opposite direction with a speed of 50.0 km/h. What is (a) the velocity of the car relative to the truck and (b) the velocity of the truck relative to the car?

13. Two canoeists in identical canoes exert the same effort paddling in a river. One paddles directly upstream (and moves upstream), whereas the other paddles directly downstream. An observer on the riverbank reckons their speeds to be 1.20 m/s and 2.90 m/s. How fast is the river flowing?

14. The pilot of an aircraft wishes to fly due west in a wind blowing at 50.0 km/h toward the south. If the speed of the aircraft in the absence of a wind is 200 km/h, (a) in what direction should the aircraft head and (b) what should its speed be relative to the ground?

15. A boy throws a ball into the air as hard as he can. He then immediately runs as fast as he can under the ball and just catches it. If his maximum speed in throwing the ball is 20.0 m/s and his best time for a 20.0-m dash is 3.00 s, how high does the ball rise?

Chapter 4

16. A cannon is fired at an angle of 30.0° above the horizontal from a cliff that is 20.0 m high. What is the initial speed of the projectile if it lands 40.0 m from the base of the cliff?

17. A particle has velocity components

$$v_x = +4 \text{ m/s} \qquad v_y = -(6.00 \text{ m/s}^2)t + 4.00 \text{ m/s}$$

Calculate the speed of the particle at $t = 2.00$ s and the direction $\theta = \tan^{-1}(v_y/v_x)$ of the velocity vector at that time.

18. A boat requires 2.00 min to cross a river that is 150 m wide. The boat's speed relative to the water is 3 m/s and the river current has a speed of 2.00 m/s. At what possible upstream or downstream points does the boat reach the opposite shore?

19. A U.S. Olympic decathlon star, who happens to be a bright physics student, is trapped on the roof of a burning building with a pencil, paper, pocket calculator, and his favorite physics textbook. He has about 15 min to decide whether to jump to the next building by either running at top speed horizontally off the edge or by using the long-jump technique. The next building is horizontally 30 ft away and vertically 10 ft below. His 100-m dash time is 10.3 s, and his long-jump distance is 25.5 ft. (Assume he long-jumps at an angle of 45° above the horizontal.) Perform calculations to decide which method (if any) he can use to reach the other building safely.

20. A projectile is fired from the origin with an initial speed of v_0 at an angle θ_0 to the horizontal. Write programs that will enable you to tabulate the projectile's x and y coordinates, displacement, x and y components of velocity, and its speed as functions of time. Tabulate the above values for the following inputs of $v_0 = 50$ m/s, $\theta_0 = 60°$ at time intervals of 0.2 s until a total time of 4.4 s is reached.

Chapter 4

21. A ball bearing is dropped from the point $x = 4$ m, $y = 2$ m. At the same moment, a second bearing is launched from $x = 0$, $y = 0$ at an angle of 20° above the positive x axis at a speed of 6 m/s. Determine (a) the minimum distance between the bearings, and (b) the time at which this minimum occurs. *Suggestions:* If you are unable to solve this problem analytically, you may wish to write and run a short computer program that locates the minimum of the *square* of the distance between the bearings.

22. A projectile is fired from the origin with an initial speed of v_0 at an angle θ_0 to the horizontal. Write programs that will enable you to tabulate the projectile's x and y coordinates, displacement, x and y components of velocity, and its speed as functions of time. Tabulate the above values for the following inputs of $v_0 = 50$ m/s, $\theta_0 = 60°$ at time intervals of 0.2 s until a total time of 4.4 s is reached.

23. A ball bearing is dropped from the point $x = 4$ m, $y = 2$ m. At the same moment, a second bearing is launched from $x = 0$, $y = 0$ at an angle of 20° above the positive x axis at a speed of 6 m/s. Determine (a) the minimum distance between the bearings, and (b) the time at which this minimum occurs. *Suggestions:* If you are unable to solve this problem analytically, you may wish to write and run a short computer program that locates the minimum of the *square* of the distance between the bearings.

Chapter 5

1. A 6.00-kg object undergoes an acceleration of 2.00 m/s^2.
(a) What is the magnitude of the resultant force acting on it?
(b) If this same force is applied to a 4.00-kg object, what acceleration does it produce?

2. A force of 10.0 N acts on a body of mass 2.00 kg. What are (a) the acceleration of the body, (b) its weight in newtons, and (c) its acceleration if the force is doubled?

3. Verify the following conversions:
(a) 1 N = 10^5 dynes, (b) 1 N = 0.225 lb.

4. On a hypothetical planet, gravitational acceleration at the surface of the planet is $0.072g$. If an object weighs 25 N on the Earth, what is its weight on the surface of the hypothetical planet?

5. What is the mass of an astronaut whose weight on the Moon is 115 N? The gravitational acceleration on the Moon is 1.63 m/s^2.

6. A net force applied to a 2.00-kg mass gives it an acceleration of 3.00 m/s^2 east. The same net force results in an acceleration of 1.00 m/s^2 east when applied to a different mass m. Determine m and the net force.

7. A 9.0-kg object undergoes an acceleration of 2.0 m/s^2 to the right under the action of two forces, $\mathbf{F}_1$ and $\mathbf{F}_2$. If $\mathbf{F}_1$ acts to the right and has a magnitude of 25 N, what are the magnitude and direction of $\mathbf{F}_2$?

Chapter 5

8. A 15-lb block rests on the floor. (a) What force does the floor exert on the block?(b) If a rope is tied to the block and run vertically over a pulley and the other end attached to a free-hanging 12-lb weight, what is the force exerted by the floor on the 15-lb block? (c) If we replace the 12-lb weight in (b) by a 25-lb weight, what is the force exerted by the floor on the 15-lb block?

9. The parachute on a race car of weight 7520 N opens at the end of a quarter-mile run when the car is traveling at 65.0 m/s What is the total retarding force required to stop the car in a distance of 955 m in the event of a brake failure?

10. A 975-kg car moving at 25.0 m/s is brought to rest in a distance of 85.0 m by a constant net force, **F**, directed opposite the velocity of the car. Determine the (a) magnitude of the acceleration of the car and (b) the magnitude of **F**.

11. Masses of 3.00 kg and 5.00 kg are connected by a massless string that passes over a frictionless pulley as in Figure P5.11. Determine (a) the tension in the string, (b) the magnitude of the acceleration of each mass, and (c) the distance each mass moves in the first second of motion if they start from rest.

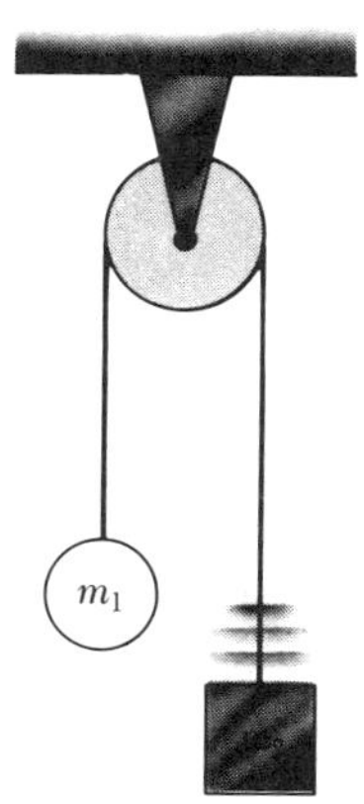

Figure P5.11

Chapter 5

12. A block is given an initial velocity of 5.00 m/s up a frictionless 20.0° incline (Fig. P5.12). How far up the incline does the block slide before stopping?

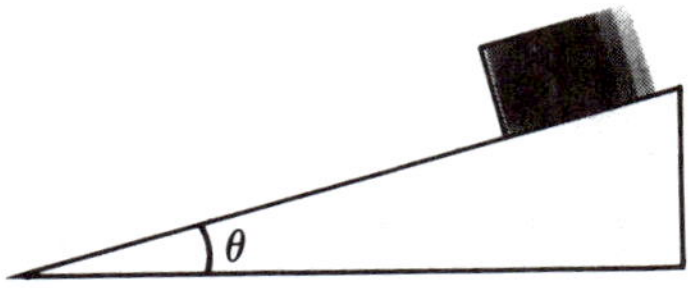

Figure P5.12

13. A 3.0-kg mass moves in a plane with its x and y coordinates given by $x = 5t^2 - 1$ and $y = 3t^3 + 2$, where x and y are in meters and t is in seconds. Find the magnitude of the net force acting on this mass at $t = 2.0$ s.

14. A child stands on the surface of a frozen pond, 12 m from shore. If the coefficient of static friction between her boots and the ice is 0.050, determine the minimum time required for her to walk to shore without slipping.

15. In order to determine the coefficients of friction between rubber and various surfaces, a student uses a rubber eraser and an incline. In one experiment the eraser slips down the incline when the angle of inclination is 46° and then moves down the incline with constant speed when the angle is reduced to 33°. From these data, determine the coefficients of static and kinetic friction for this experiment.

16. A racing car accelerates uniformly from 0 to 80.0 mi/h in 8.00 s. The external force that accelerates the car is the frictional force between tires and road. If the tires do not spin, determine the *minimum* coefficient of friction.

Chapter 5

17. The largest caliber antiaircraft gun operated by the Luftwaffe during World War II was the 12.8-cm Flak 40. This weapon fired a 25.8-kg shell with a muzzle speed of 880 m/s. What propulsive force was necessary to attain the muzzle speed within the 6.0-m barrel? (Assume constant acceleration and neglect the Earth's gravitational effect.)

18. Three blocks are in contact with each other on a frictionless, horizontal surface as in Figure P5.18. A horizontal force **F** is applied to m_1. If $m_1 = 2.00$ kg, $m_2 = 3.00$ kg, $m_3 = 4.00$ kg, and $F = 18.0$ N, find (a) the magnitude of the acceleration of the blocks, (b) the net force on each block, and (c) the magnitude of the contact forces between the blocks.

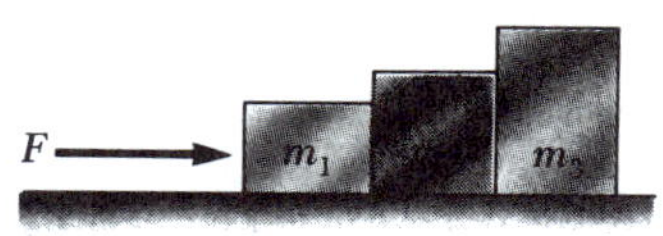

Figure P5.18

19. Repeat Problem 69 for the case where the surface is rough and the coefficient of kinetic friction between blocks and surface is 0.10.

20. Two masses m and M are attached with strings as shown in Figure P5.20. If the system is in equilibrium, show that $\tan\theta = 1 + \frac{2M}{m}$.

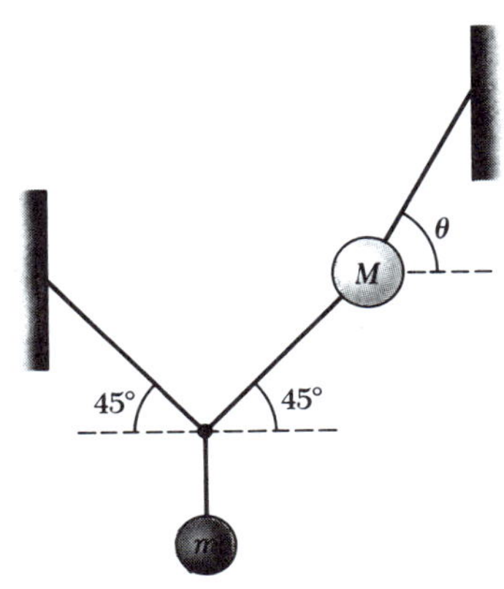

Figure P5.20

Chapter 6

1. What centripetal force is required to keep a 1.50 kg mass moving in a circle of radius 0.400 m at a speed of 4.00 m/s?

2. A 1500-kg car rounds an unbanked curve with a radius of 52.0 m at a speed of 12.0 m/s. What minimum coefficient of friction must exist between road and tires to prevent the car from slipping?

3. A Ferris wheel with radius 20.0 m makes 1 rev every 9.00 s. What force does a 55.0-kg passenger exert on the seat when at the top of the Ferris wheel?

4. A ball is suspended from the ceiling of a moving car by a string 25.0 cm in length. An observer in the car notes that the ball deflects 6.00 cm from the vertical toward the rear of the car. What is the acceleration of the car?

5. An 80.0 kg skydiver jumps from a slow-moving aircraft and reaches a terminal speed of 50.0 m/s. (a) What is the acceleration of the skydiver when her speed is 30.0 m/s? What is the resistive force on her when her speed is (b) 50.0 m/s and (c) 30.0 m/s?

6. (a) Estimate the terminal speed of a wooden sphere (density 0.830 g/cm^3) moving in air if its radius is 8.00 cm. (b) From what height would a freely falling object reach this speed in the absence of air resistance?

7. A spinning ball of radius 5.00 cm slows uniformly from 30.0 rev/min to rest in 0.300 s. Compute the radial, tangential, and net acceleration of a point on the equator of the ball at the beginning of this time period.

Chapter 6

8. A 0.400-kg pendulum bob passes through the lowest part of its path with a speed of 8.20 m/s. What is the tension in the cable if the it is 80.0 cm long?

9. A small turtle, appropriately named ``Dizzy,'' is placed on a horizontal, rotating turntable 20.0 cm from its center. Dizzy's mass is 50.0 g, and the coefficient of static friction between his feet and the turntable is 0.300. Find (a) the maximum number of revolutions per second the turntable can have if Dizzy is to remain stationary relative to it and (b) Dizzy's speed and radial acceleration when he is on the verge of slipping.

10. A railroad track has a curve of 400-m radius. The track is banked toward the inside at an angle of 6.00°. For trains of what speed was this track designed? (Assume that the correct speed requires only the normal force to keep the train on the track.)

11. A 4.00-kg mass is attached to a horizontal rod by two strings, as in Figure P6.11. The strings are under tension when the rod rotates about its axis. If the speed of the mass is 4.00 m/s when observed at the following positions, find the tension in the string when the mass is (a) at its lowest point, (b) in the horizontal position, and (c) at its highest point.

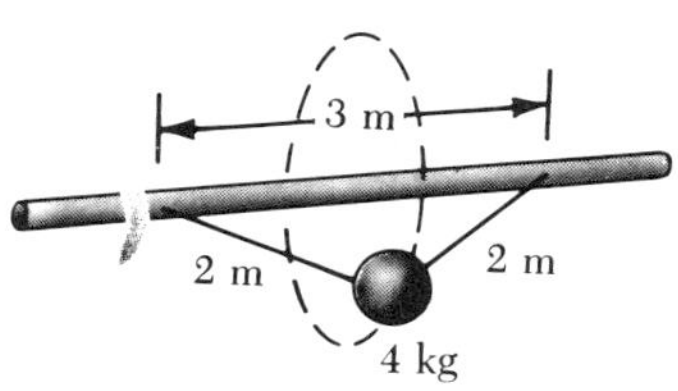

Figure P6.11

12. Suppose the rod in Figure P6.11 is made vertical and rotates about this axis. If the mass rotates at a constant speed of 6.00 m/s in a horizontal plane, determine the tensions in the upper and lower strings.

Chapter 7

1. A 65.0-kg woman climbs a flight of 20 stairs, each 23.0 cm high. How much work is done against the force of gravity?

2. A tugboat exerts a constant force of 5000 N on a ship moving at constant speed through a harbor. How much work does the tugboat do on the ship in 3.00 km?

3. Verify the following energy unit conversions:
(a) $1\text{ J} = 10^7$ ergs, (b) $1\text{ J} = 0.737$ ft·lb.

4. For $\mathbf{A} = -2\mathbf{i} + 3\mathbf{j}$, find (a) the magnitude of **A** and (b) the angle that **A** makes with the positive y axis. [In (b), use the definition of the scalar product.]

5. As a particle moves from the origin to $(3\mathbf{i} - 4\mathbf{j})$ m, it is acted upon by a force given by $(4\mathbf{i} - 5\mathbf{j})$ N. Calculate the work done by this force as the particle moves through the given displacement.

6. An object moves from $x = 0$ to $x = 3.00$ m. If the resultant force acting on that object is in the x direction and varies as shown in Figure 7.14, determine the total work done on the object.

7. The force acting on a particle is given by $F_x = (8.0x - 16)$ N, where x is in meters. (a) Make a plot of this force versus x from $x = 0$ to $x = 3.0$ m. (b) From your graph, find the net work done by this force as the particle moves from $x = 0$ to $x = 3.0$ m.

8. If an applied force varies with position according to $F_x = 3.0x^3 - 5.0$, where x is in meters, how much work is done by this force on an object that moves from $x = 4.0$ m to $x = 7.0$ m?

Chapter 7

9. When a 4.00-kg mass is hung vertically on a certain light spring that obeys Hooke's law, the spring stretches 2.50 cm. If the mass is removed, (a) how far will the spring stretch if a 1.50-kg mass is hung on it? (b) How much work must an external agent do to stretch the same spring 4.00 cm from its unstretched position?

10. A 0.30-kg ball has a speed of 15 m/s. (a) What is its kinetic energy? (b) If its speed is doubled, what is its kinetic energy?

11. Calculate the kinetic energy of a 1000-kg satellite orbiting the Earth at a speed of 7.00×10^3 m/s.

12. A pole-vaulter clears a height of 6.00 m. With what speed does he strike the mat in the landing area?

13. ``Winkin, Blinkin and Nod one day . . ." abandoned their boat and took to the slopes, complete with sled. They found a hill of slope 2.87° and 30.0 m long. Starting from rest at the top of the hill, they slid to the bottom, where their speed was 5.00 m/s. If the combined mass of the three plus sled was 70.0 kg, what was the average frictional force on the slope?

14. In the giant roller coaster in Gurnee, Illinois (see Problem 22, Chapter 6), the speed of the cars is 13 m/s at the top of a loop that is 40 m in height. What is their speed at the bottom if we neglect any friction effects?

15. A weight lifter lifts 250 kg through 2.00 m in 1.50 s. What is his power output?

Chapter 7

16. A 65.0-kg athlete runs 600 m up a mountain inclined at 20.0° to the horizontal. He performs this feat in 80.0 s. Neglecting air resistance, (a) how much work does he perform and (b) what is his power output during the run?

17. The car described in Table 7.4 travels at 35.9 m/s. Determine (a) the power needed to overcome air drag, and (b) the total power delivered to the wheels.

18. A 15-kg box is dragged at uniform speed up an incline that is 8.00 m long and makes a 15° angle with the horizontal. If the coefficient of friction between box and incline is 0.40, how much work is done by (a) the applied force, (b) the normal force, and (c) the gravitational force? (d) How much energy is lost due to friction?

19. A woman raises a 10-kg flag from the ground to the top of a 10-m flagpole at 0.25 m/s. What work is done by (a) the woman, (b) gravity? (c) What is the power output of the woman?

20. The force $F_x(x)$ given in Figure P7.20 acts on an object as the object moves along the x axis. Use graphical techniques to estimate the work done by this force as the object moves from $x = 0$ to $x = 8.00$ m.

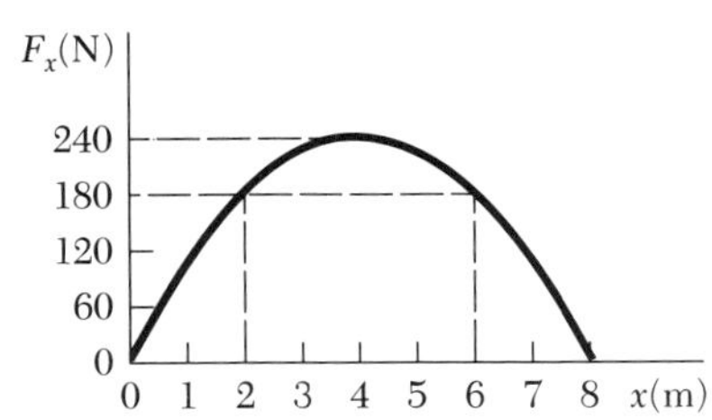

Figure P7.20

Chapter 7

21. A rope tow pulling skiers up a 30° slope 600 m long moves at 3.0 m/s and carries a maximum of 120 passengers. The average mass of each passenger is 80 kg. Neglecting friction, determine the power rating a motor must have in order to operate the tow under maximum load conditions.

22. Referring to Problem 83, find (a) the instantaneous rate at which work is being done on the projectile and (b) if the mass of the projectile is 10.0 kg and the initial height is 40 m, the instantaneous rate at which work is being done after 1.00 s, 2.00 s, and 3.00 s. (*Note:* Be careful of the elapsed time.)

23. A 5.00-kg particle starts at the origin and moves along the x axis. The net force acting on the particle is measured at intervals of 1.00 m to be 27.0, 28.3, 36.9, 34.0, 34.5, 34.5, 46.9, 48.2, 50.0, 63.5, 13.6, 12.2, 32.7, 46.6, and 27.9 N. Determine the total work done on the particle over this interval.

Chapter 8

1. A particle moves in the xy plane in Figure P8.1 under the influence of a frictional force that opposes its displacement. If the frictional force has a magnitude of 3 N, calculate the total work done by friction along the following *closed* paths: (a) the path OA followed by the return path AO, (b) the path OA followed by AC and the return path CO, and (c) the path OC followed by the return path CO. (d) Your results for the three closed paths should all be different and nonzero. What is the significance of this?

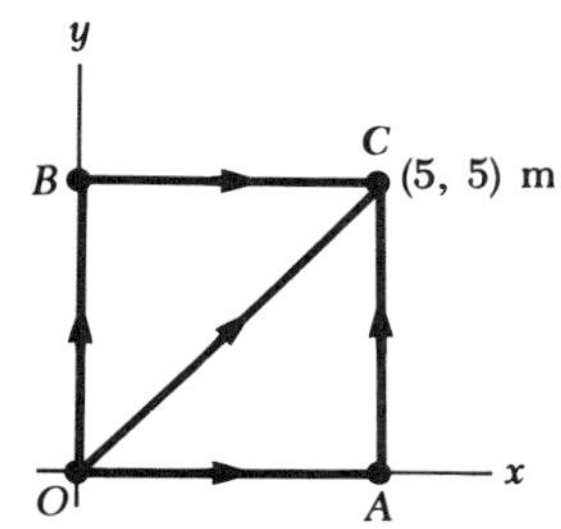

Figure P8.1

2. A 5.0-kg mass is attached to a light string of length 2.0 m to form a pendulum (Fig. P8.2). The mass is given an initial speed of 4.0 m/s at its lowest position. When the string makes an angle of 37° with the vertical, find (a) the change in the potential energy of the mass, (b) the speed of the mass, and (c) the tension in the string. (d) What is the maximum height reached by the mass above its lowest position?

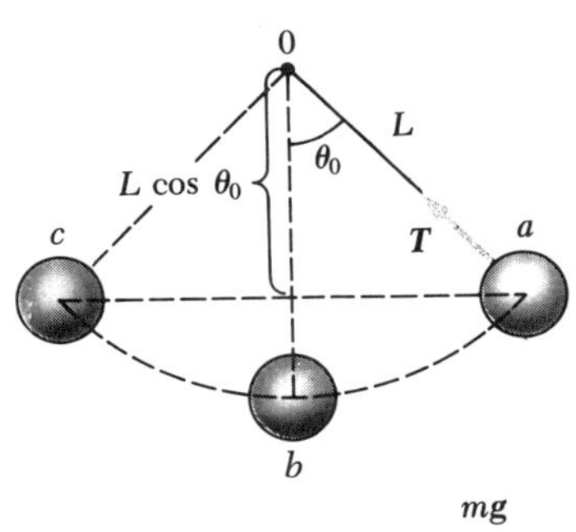

Figure P8.2

Chapter 8

3. A 0.200-kg ball is thrown vertically upward from the ground with an initial speed of 20.0 m/s. Let y measure the height of the ball above the ground, and choose the gravitational potential energy of the ball to be zero at the ground. Write an expression for (a) the gravitational potential energy as a function of y, (b) the total mechanical energy as a function of y, and (c) the kinetic energy of the ball as a function of y. For what range of values of y is this expression valid? (d) What is the maximum height does the ball attain?

4. A 0.400-kg ball is thrown into the air and reaches a maximum altitude of 20.0 m. Taking its initial position as the point of zero potential energy and using energy methods, find (a) its initial speed, (b) its total mechanical energy, and (c) the ratio of its kinetic energy to its potential energy when its altitude is 10.0 m.

5. A 25.0-kg child on a swing 2.00 m long is released from rest when the swing supports make an angle of 30.0° with the vertical. (a) Neglecting friction, find the child's speed at the lowest position. (b) If the speed of the child at the lowest position is 2.00 m/s, what is the energy loss due to friction?

6. A force F_x, shown as a function of distance in Figure P8.6, acts on a 5.0-kg mass. If the mass starts from rest at $x = 0$ m, determine its speed at $x = 2.0$, 4.0, and 6.0 m.

Figure P8.6

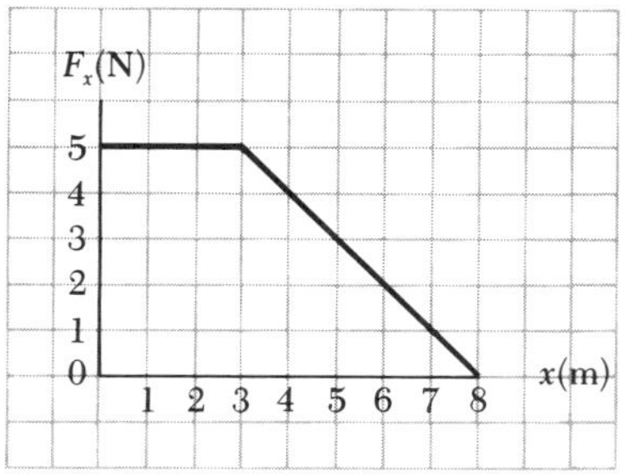

Chapter 8

7. A mass of 2.5 kg is attached to a spring that has a constant $k = 65$ N/m. The spring is stretched and allowed to oscillate such that the mass moves freely on a frictionlessl, horizontal surface. When the spring is stretched 10 cm, its elastic potential energy and the kinetic energy of the mass are equal. What is the maximum speed of the mass?

8. A spring has a force constant of 300 N/m. How much work must be done on the spring to stretch it (a) 4.00 cm from its equilibrium position and (b) from $x = 2.00$ cm to $x = 4.00$ cm, where $x = 0$ is its equilibrium position? (In the unstretched position, the potential energy is defined to be zero.)

9. A spring has a force constant of 500 N/m. What is the elastic potential energy stored in it when (a) it is stretched 4.00 cm from equilibrium, (b) it is compressed 3.00 cm from equilibrium, and (c) it is unstretched?

10. The potential energy function for a system is $U = ax^2 - bx$, where a and b are constants. (a) Find the force $\mathbf{F}_x$ associated with this potential energy function. (b) At what value of x is the force zero?

11. An olympic high jumper who is 2.0 m tall makes a leap of 2.3 m over a horizontal bar. Estimate the speed with which he must leave the ground to perform this feat. (*Hint:* Estimate the position of the center of his body before jumping, and assume he is in a horizontal position when he reaches the peak of his jump.)

Chapter 8

12. A potential energy function for a system is given by $U(x) = -x^3 + 2x^2 + 3x$. (a) Determine the force $\mathbf{F}_x$ as a function of x. (b) For what values of x is the force equal to zero? (c) Plot $U(x)$ versus x and F_x versus x and indicate points of stable and unstable equilibrium.

13. A crane is to lift 2000 kg of material to a height of 150 m in 1.0 min at a uniform rate. What electric power is required to drive the crane motor if 35 percent of the electric power is converted to mechanical power? (Is it reasonable to neglect the kinetic energy?).

14. An 80 000-kg airliner is flying at 900 km/h at a height of 10.0 km. What is its total energy (kinetic plus potential)?

15. The potential energy function of a particle as a function of distance x is $U = Ax^2 - Bx^3$, where A and B are positive constants. (a) Find the expression for the force the particle experiences as a function of x. (b) Locate the point(s) where the force vanishes. (c) Is the equilibrium of the particle stable at these points? (d) What sort of motion results when the particle is released from rest a small distance from each point where the force is zero?

Chapter 9

1. The momentum of a 1250-kg car is equal to the momentum of a 5000-kg truck traveling at a speed of 10.0 m/s. What is the speed of the car?

2. Calculate the magnitude of the linear momentum for the following cases: (a) a proton of mass 1.67×10^{-27} kg moving with a speed of 5.00×10^{6} m/s; (b) a 15-g bullet moving with a speed of 500 m/s; (c) a 75-kg sprinter running at a speed of 12 m/s, and (d) the Earth (mass 5.98×10^{24} kg) moving with an orbital speed of 2.98×10^{4} m/s.

3. A 40.0-kg child standing on a frozen pond throws a 2.00-kg stone to the east with a speed of 8.00 m/s. Neglecting friction between child and ice, find the recoil velocity of the child.

4. A 60.0-kg boy and a 40.0-kg girl, both wearing skates, face each other at rest. The boy pushes the girl, sending her eastward with a speed of 4.00 m/s. Describe the subsequent motion of the boy. (Neglect friction.)

5. A 2.5-kg mass moving initially with a speed of 10 m/s makes a perfectly inelastic, head-on collision with a 5.0-kg mass initially at rest. (a) Find the final speed of the composite particle. (b) How much energy is lost in the collision?

6. Consider the ballistic pendulum described in Example 9.10 of the text and shown in Figure 9.11. (a) Show that the ratio of the kinetic energy after the collision to the kinetic energy before the collision is equal to $m_1/(m_1 + m_2)$, where m_1 is the mass of the bullet and m_2 is the mass of the block. (b) If $m_1 = 8.00$ g and $m_2 = 2.00$ kg, what percentage of the original energy is left after the inelastic collision? What accounts for the missing energy?

Chapter 9

7. A bullet (m = 0.010 kg) moving at 300 m/s strikes a wooden fence 2.0 cm thick and comes out the other side at a speed of 200 m/s. Assuming constant acceleration while passing through the fence, determine the force experienced by the fence during contact with the bullet.

8. Two billiard balls have velocities of 2.00 m/s and - 0.500 m/s before they meet in an elastic head-on collision. What are their final velocities?

9. A golf club of mass 200 g is traveling at 55 m/s just before it strikes a 46-g golf ball at rest on a tee. After the collision, the clubhead travels (in the same direction) at 40 m/s. Find the speed of the golf ball just after impact.

10. How many head-on collisions will the neutron in Prob. 9.29 need to make to reduce its energy to 1.6×10^{-19} J? How will the number of head-on collisions change if carbon is replaced by an element whose nucleus is one third as massive as the carbon nucleus?

11. Two particles of masses m and $3m$ are moving toward each other along the x direction with the same initial speeds v_0. They undergo a head-on elastic collision and each rebounds along the same line as it came in on. Find the final speeds of the particles.

12. A proton moving with a velocity $v_0\mathbf{i}$ collides elastically with a proton initially at rest. If both protons have the same speed after the collision, find (a) the speed of each proton after the collision in terms of v_0 and (b) the direction of the velocity vectors after the collision.

Chapter 9

13. A 10-kg mass initially at rest explodes into three pieces. A 4.5-kg piece goes north at 20 m/s, and a 2.0-kg piece moves eastward at 60 m/s. (a) Determine the velocity of the third piece. (b) Find the energy of the explosion.

14. Three masses located in the *xy* plane have the following coordinates: a 3.00-kg mass, (3.00, - 2.00) m; a 4.00-kg mass, (- 2.00, 4.00) m; a 1.00-kg mass, (2.00, 2.00) m. Find the coordinates of the center of mass of the three-particle system.

15. A shell is fired from a cannon at level ground to hit an enemy bunker situated a distance R away from the cannon. At the highest point of the trajectory, the shell explodes into two equal parts. One part hits the ground at $R/2$. At what distance from the cannon does the second part fall?

16. Compute the location of the center of mass of a piece of plywood cut into the shape of an isosceles right triangle if the two equal sides are 1.3 m long. The triangle is oriented such that the x axis bisects the 90° angle of the triangle. (*Hint:* Use the integral form of the center-of-mass equation.)

17. A 5.00-kg particle moves along the x axis with a velocity of + 3.00 m/s. A 2.00-kg particle moves along the x axis with a velocity of - 2.50 m/s. Find (a) the velocity of the center of mass and (b) the total momentum of the system.

18. A large rocket with an exhaust speed v_e = 3000 m/s develops a thrust of 2.4×10^7 N. (a) How much mass is being blasted out of the rocket exhaust per second? (b) What is the maximum speed the rocket can acquire if it starts from rest in a force-free environment with v_e = 3.0 km/s and 90 percent of its initial mass is fuel?

Chapter 9

19. Consider a sphere of radius R and mass density ρ that is solid except for a spherical hollow volume of radius $R/2$. The center of the spherical void is located at a distance $R/2$ from the center of the sphere. Find the center of mass of the body. (*Hint:* Treat the void as a negative mass.)

20. A loaded howitzer is rigidly bolted to a railroad flatcar. The mass of the combination is 20 000 kg, and the flatcar coasts along a level, straight, frictionless track with a constant speed V. The howitzer is set to an elevation angle of 60° with the horizontal and then fired. The flatcar and attached howitzer recoil, moving in the opposite direction along the track with the same speed V. (a) If the muzzle speed of the howitzer shell is 300 m/s, and if the shell has a mass of 50 kg, find V. (b) How high does this shell rise in the air?

21. A spacecraft is stationary in deep space when its rocket engine is ignited for a 100-s "burn." Hot gases are ejected at a constant rate of 150 kg/s with a speed relative to the spacecraft of 3000 m/s. The initial mass of the spacecraft (plus fuel and oxidizer) is 25 000 kg. Determine the thrust of the rocket engine and the initial acceleration in units of g. What is the final speed of the spacecraft?

22. Consider a head-on elastic collision between a moving particle of mass m_1 and an initially stationary particle of mass m_2 (see Example 9.12). (a) Plot f_2, the fraction of energy transferred to m_2, as a function of the ratio m_2/m_1 and show that f_2 reaches a maximum when $m_2/m_1 = 1$. (b) Perform an analytical calculation that verifies that f_2 is a maximum when $m_1 = m_2$.

Chapter 10

1. A wheel rotates in such a way that its angular displacement in a time t is given by $q = at^2 + bt^3$, where a and b are constants. Determine equations for (a) the angular speed and (b) the magnitude of its angular acceleration, both as functions of time.

2. A grinding wheel, initially at rest, is rotated with constant angular acceleration $a = 5.00\ \text{rad/s}^2$ for 8.00 s. It is then brought to rest with uniform negative acceleration in 10.0 rev. Determine the negative acceleration required and the time needed to bring the wheel to rest.

3. An automobile accelerates from zero to 30 m/s in 6.0 s. The wheels have a diameter of 0.40 m. What is the magnitude of the angular acceleration of each wheel?

4. A tire of moment of inertia $80.0\ \text{kg}\bullet\text{m}^2$ rotates about a fixed central axis at the rate of 600 rev/min. What is its kinetic energy?

5. Three particles each of mass M are arranged at the vertices of an equilateral triangle as shown in Figure P10.5. Determine the moments of inertia about the x, y and z axes. The z axis passes through O and is normal to the x-y plane.

Figure P10.5

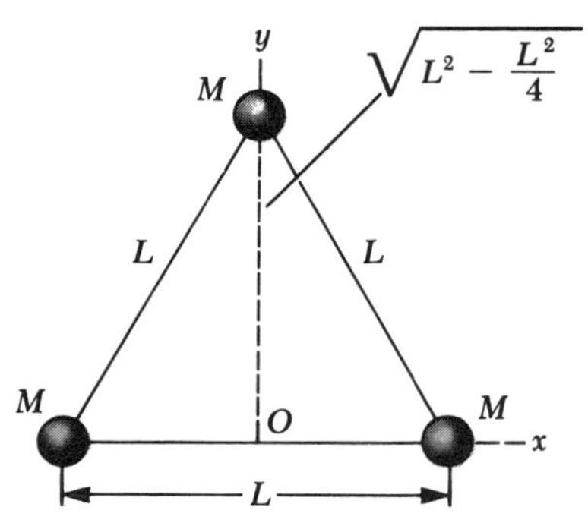

Chapter 10

6. Following the procedure used in Example 10.6, prove that the moment of inertia about the y' axis of the rigid rod in Figure 10.9 is $\frac{1}{3}ML^2$.

7. Calculate the net torque (magnitude and direction) on the beam shown in Figure P10.7 about (a) an axis through O perpendicular to the plane of the page and (b) an axis through C perpendicular to the plane of the page.

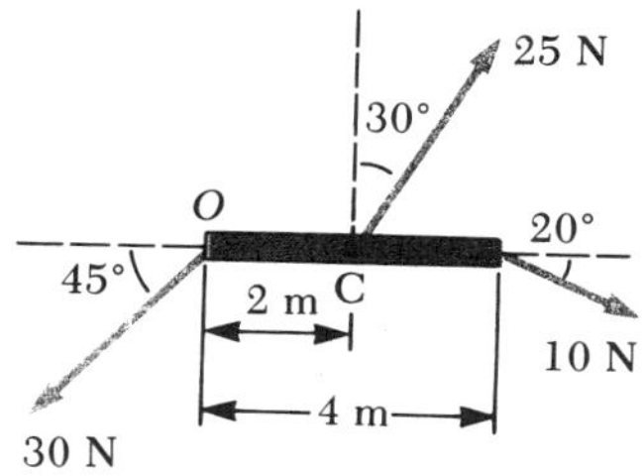

Figure P10.7

8. The combination of an applied force and a frictional force produces a constant total torque of 36 N•m on a wheel rotating about a fixed axis. The applied force acts for 6 s, during which time the angular speed of the wheel increases from 0 to 10 rad/s. The applied force is then removed, and the wheel comes to rest in 60 s. Find (a) the moment of inertia of the wheel, (b) the magnitude of the frictional torque, and (c) the total number of revolutions of the wheel.

9. If a motor is to produce a torque of 50 N•m on a wheel rotating at 2400 rev/min, how much power must the motor deliver?

Chapter 10

10. A wheel 1.00 m in diameter rotates on a fixed, frictionless, horizontal axle. Its moment of inertia about this axis is $5.00 \text{ kg} \bullet \text{m}^2$. A constant tension of 20.0 N is maintained on a rope wrapped around the rim of the wheel, so as to cause the wheel to accelerate. If the wheel starts from rest at $t = 0$, find (a) its angular acceleration, (b) its angular speed at $t = 3.00$ s, (c) its kinetic energy at $t = 3.00$ s, and (d) the length of rope unwound in the first 3.00 s.

11. Calculate the moment of inertia of a uniform solid sphere of mass M and radius R about a diameter (see Table 10.2). (*Hint*: Treat the sphere as a set of disks of various radii, and first obtain an expression for the moment of inertia of one of these disks about the symmetry axis.)

12. A uniform solid cylinder of mass M and radius R rotates on a horizontal, frictionless axle (Fig. P10.12). Two equal masses hang from light cords wrapped around the cylinder. If the system is released from rest, find (a) the tension in each cord, (b) the magnitude of the acceleration of each mass, and (c) the angular speed of the cylinder after the masses have descended a distance h..

Figure P10.12

Chapter 11

1. The centers of mass of a solid sphere and a solid cylinder, each of mass M and radius R, are moving with speed v relative to the floor. Find the ratios of their total kinetic energies.

2. Given $\mathbf{M} = 6\mathbf{i} + 2\mathbf{j} - \mathbf{k}$ and $\mathbf{N} = 2\mathbf{i} - \mathbf{j} - 3\mathbf{k}$, calculate the vector product $\mathbf{M} \times \mathbf{N}$.

3. Two forces $\mathbf{F_1}$ and $\mathbf{F_2}$ act along the two sides of an equilateral triangle as shown in Figure P11.3. Find a third force $\mathbf{F_3}$ to be applied at B and along BC that will make the net torque about the point of intersection of the altitudes zero. Will the net torque change if $\mathbf{F_3}$ is applied not at B but at any other point along BC?

Figure P11.3

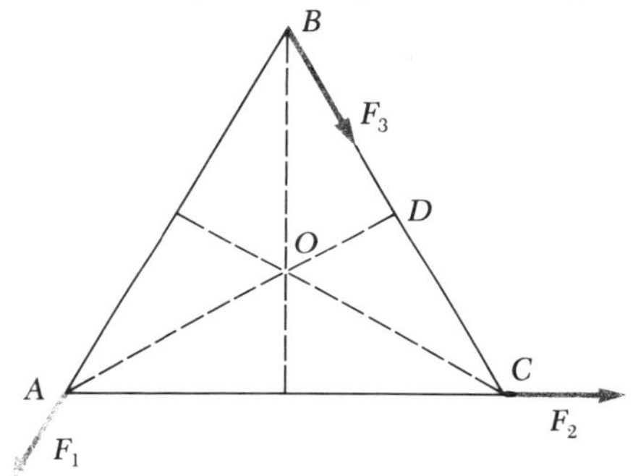

4. Two particles move in opposite directions along a straight line (Fig. P11.4). The particle of mass m moves to the right with a speed v , while the particle of mass $3m$ moves to the left with a speed v. What is the total angular momentum of the system relative to (a) the point A, (b) the point O, and (c) the point B?

Figure P11.4

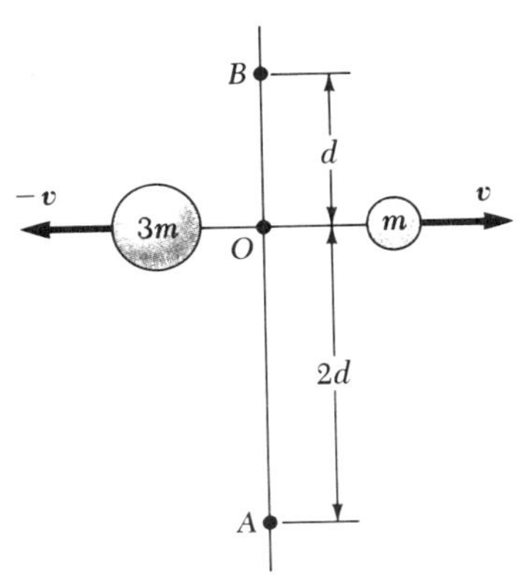

Chapter 11

5. An airplane of mass 12 000 kg flies level to the ground at an altitude of 10.0 km with a constant speed of 175 m/s relative to the Earth. (a) What is the magnitude of the airplane's angular momentum relative to a ground observer directly below the airplane? (b) Does this value change as the airplane continues its motion along a straight line?

6. A 4.0-kg mass is attached to a light cord, which is wound around a pulley (Fig. 10.18). The pulley is a uniform solid cylinder of radius 8.0 cm and mass 2.0 kg. (a) What is the net torque on the system about the point O? (b) When the mass has a speed v, the pulley has an angular speed $\omega = v/R$. Determine the total angular momentum of the system about O. (c) Using the fact that $\boldsymbol{\tau} = d\mathbf{L}/dt$ and your result from part (b), calculate the acceleration of the mass.

7. Two point masses, each 4.0 kg, are joined by a light rigid rod 0.40 m long. This ``dumbbell'' system is placed on a horizontal frictionless surface and set into rotation about a vertical axis through the center of mass with an angular speed of 3.0 rad/s. Find (a) the moment of inertia of the system about the center of mass and (b) the angular momentum about the center of mass. (c) What tangential force, applied to each mass, will produce an angular acceleration of magnitude 6.0 rad/s^2?

Chapter 11

8. A particle of mass $m = 10.0$ g and speed $v_0 = 5.00$ m/s collides with and sticks to the edge of a uniform solid sphere of mass $M = 1.00$ kg and radius $R = 20.0$ cm (Fig. P11.8). If the sphere is initially at rest and is pivoted about a frictionless axle through O perpendicular to the plane of the paper, (a) find the angular speed of the system after the collision and (b) determine how much energy is lost in the collision.

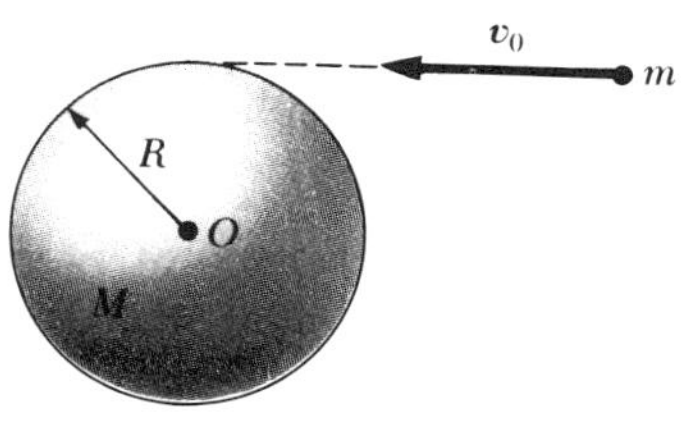

Figure P11.8

9. Consider Problem 33 problem with $\ell = 2.00$ m, $M = 2.00$ kg, $m = 10.0$ g, and $v = 200$ m/s. In this case the bullet passes all the way through the block, exiting with speed $v = 25.0$ m/s horizontal to the surface. (a) Determine the angular momentum of the block just after the bullet exits. (b) Determine the kinetic energy lost in this collision. Neglect the transit time through the block.

10. Consider the problem of the solid sphere rolling down an incline as described in Example 11.1. (a) Choose instantaneous axis through the contact point P as the axis of the origin for the torque equation and show that the acceleration of the center of mass is $a_{CM} = \frac{5}{7}g \sin\theta$. . (b) Show that the minimum coefficient of friction such that the sphere rolls without slipping is $\mu_{min} = \frac{2}{7}\tan\theta$..

11. The position vector of a 5.0-kg particle is $\mathbf{r} = (2.0t^2\mathbf{i} + 3.0\mathbf{j})$ m, where t is in seconds. Determine the angular momentum and the torque acting on the particle about the origin.

Chapter 11

12. A particle of mass m is located at the vector position **r** and has a linear momentum **p**. (a) If **r** and **p** both have nonzero x, y, and z components, show that the angular momentum of the particle relative to the origin has components $L_x = yp_z - zp_y$, $L_y = zp_x - xp_z$, and $L_z = xp_y - yp_x$. (b) If the particle moves only in the xy plane, prove that $L_x = L_y = 0$ and $L_z = 0$.

13. A large, cylindrical roll of tissue paper of initial radius R lies on a long, horizontal surface with the open end of the paper nailed to the surface. The roll is given a *slight* shove ($v_0 \approx 0$) and commences to unroll. (a) Determine the speed of the center of mass of the roll when its radius has diminished to r. (b) Calculate a numerical value for this speed at r = 1.0 mm, assuming R = 6.0 m. (c) What happens to the energy of the system when the paper is completely unrolled? (*Hint*: Assume the roll has a uniform density and apply energy methods.)

Chapter 12

1. With reference to Figure P12.1, find x such that the normal force at O will be one half the normal force at P. Neglect the weight of the beam.

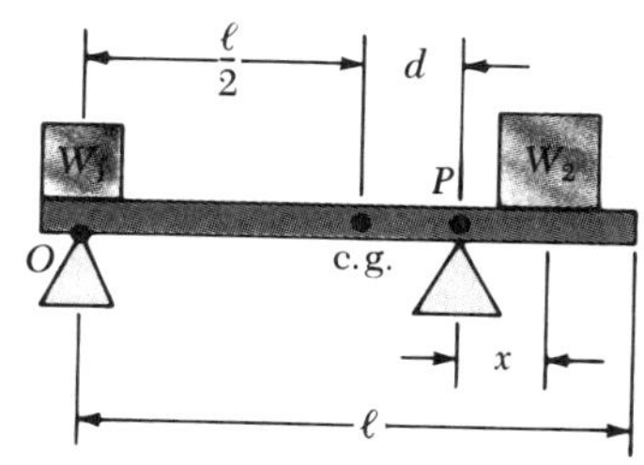

Figure P12.1

2. A flat plate in the shape of a letter T is cut with the dimensions shown in Figure P12.2. Locate the center of gravity. (Hint: Note that the weights of the two rectangular parts are proportional to their volumes.)

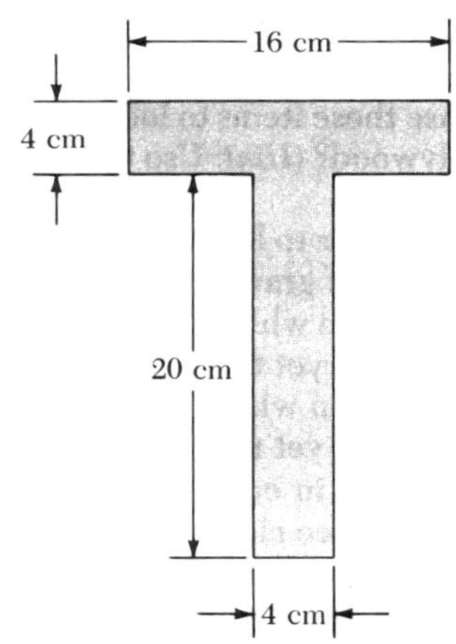

Figure P12.2

3. A rod of length L has a linear mass density $\lambda(x) = Ax$ kg/m, where A is a constant and x is measured in meters from one end of the rod. Assuming a uniform gravitational field, determine the location of the center of gravity of the rod.

4. A meter stick supported at the 50.0-cm mark has masses of 300 g and 200 g hanging at the 10.0-cm and 60.0-cm marks, respectively. Determine the position at which one would hang a 400-g mass to keep the meter stick balanced.

Chapter 12

5. A 48 kg diver stands at the end of a 3.0-m-long diving board. What torque does the weight of the diver produce about an axis perpendicular to and in the plane of the diving board through its midpoint?

6. A 10-m-long 100 kg uniform beam is supported at the two ends by two strings to hang horizontally. If one of the strings can withstand a maximum load of 600 N, at what maximum distance, measured from the stronger string, could a mass of 50 kg be placed so that the weaker string would not break?

7. A cube of steel 4.0 cm on an edge is subjected to a shearing force of 3000 N while one face is clamped. Find the shearing strain in the cube.

8. What increase in pressure is necessary to decrease the volume of a 4.0-cm-diameter sphere of mercury by 0.1 percent?

9. At a depth of 400 km below the Earth's surface, the pressure is 1.36×10^{10} Pa. If a cubic centimeter of copper were brought to the surface from this depth, what would be its new volume?

10. A solid sphere of radius R and mass M is placed in a wedge as shown in Figure P12.10. The inner surfaces of the wedge are frictionless. Determine the forces exerted by the wedge on the sphere at the two contact points.

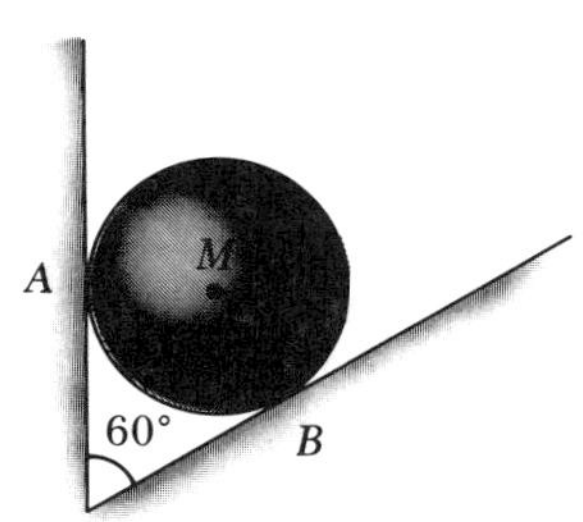

Figure P12.10

Chapter 12

11. A uniform beam of length 4.0 m and mass 10 kg supports a 20 kg mass as in Figure P12.11. (a) Draw a free-body diagram for the beam. (b) Determine the tension in the supporting wire and the components of the reaction force at the pivot.

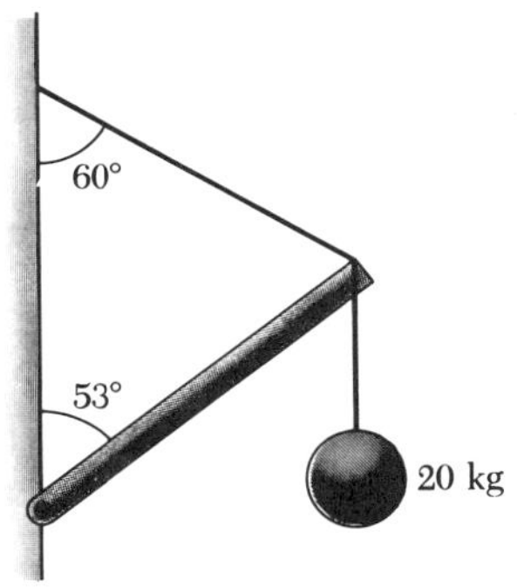

Figure P12.11

12. Two sheets of metal are joined together by four steel rivets, each 7.0 mm in diameter. Find the maximum force that can be applied to pull the sheets apart tangentially if the shearing stress on the rivets is not exceed 8.0×10^7 N/m^2. Assume that the stress is shared equally by the rivets.

13. The distortion of the Earth's crustal plates is an example of shear on a large scale. A particular type of crustal rock is determined to have a shear modulus of 1.5×10^{10} N/m^2. What shear stress is involved when a 10 km layer of rock is sheared through a distance of 5.0 m?

14. A 30 kg hammer moving at 20 m/s strikes a 2.3-cm-diameter steel spike and then rebounds with a speed of 10 m/s in 0.11 s. What is the average strain in the spike during the impact?

Chapter 13

1. For the particle described in Problem 1, determine (a) the velocity at any time t, (b) the acceleration at any time, (c) the maximum velocity and maximum acceleration, and (d) the velocity and acceleration at $t = 0$.

2. A particle oscillates with simple harmonic motion such that its displacement varies according to the expression $x = (5.0 \text{ cm}) \cos(2.0t + \pi/6)$, where x is in centimeters and t is in seconds. At $t = 0$, find (a) the displacement of the particle, (b) its velocity, and (c) its acceleration. (d) Find the period and amplitude of the motion.

3. A particle moving in simple harmonic motion travels 20.0 cm in each cycle of its motion, and its maximum acceleration is 50.0 m/s^2. Find (a) the angular frequency of the motion and (b) the maximum speed of the particle.

4. A 0.20 N weight is hung from a spring that has a force constant $k = 6.0$ N/m. How much is the spring displaced?

5. A mass-spring system oscillates such that the displacement is given by $x = (0.25 \text{ m}) \cos(2.0t)$. (a) Find the speed and acceleration of the mass when $x = 0.10$ m. (b) Determine the maximum speed and maximum acceleration.

6. A 100-g mass hangs on the end of a vertical Hooke's-law spring. When 40 g is added, the spring stretches an additional 5.0 cm. With this extra mass attached, the spring is set into vertical oscillation with an amplitude of 10 cm. (a) Find the frequency of the motion. (b) How long does the mass take to travel from the mid-position to maximum displacement? (c) Find the net force on the mass when it is at maximum upward displacement.

Chapter 13

7. A 200-g mass is attached to a spring and executes simple harmonic motion with a period of 0.25 s. If the total energy of the system is 2.0 J, find (a) the force constant of the spring and (b) the amplitude of the motion.

8. A mass-spring system oscillates with an amplitude of 3.50 cm. If the spring constant is 250 N/m and the mass is 0.500 kg, determine (a) the mechanical energy of the system, (b) the maximum speed of the mass, and (c) the maximum acceleration.

9. The velocity of a 300-g mass attached to the end of a spring is represented by $v = (1.60\ \text{m/s})\sin\omega t$, with $\omega = 2.83$ rad/s. Determine the total energy of the system.

10. Calculate the frequency and period of a simple pendulum of length 10.0 m.

11. A visitor to a lighthouse wishes to determine the height of the tower. She has a spool of thread that she uses to support a small rock as a simple pendulum hanging down the center of the spiral staircase of the tower. The period of oscillation is 9.40 s. What is the height (in meters) of the tower?

12. A uniform rod is pivoted at one end as in Figure 13.12. If the rod swings in simple harmonic motion, what must its length be in order that its period be equal to that of a simple pendulum 1.00 m long?

13. A circular hoop of radius R is hung over a knife edge. Show that its period of oscillation is equal to that of a simple pendulum of length $2R$.

14. Show that the damping constant, b, has units of kilograms per second.

Chapter 13

15. Show that Equation 13.30 is a solution to Equation 13.29 provided that $b^2 < 4mk$.

16. Consider an undamped forced oscillator, and show that Equation 13.33 is a solution of Equation 13.32, with an amplitude given by Equation 13.34.

17. A mass M is connected to a spring of mass m and oscillates in simple harmonic motion on a horizontal, smooth surface (Fig. P13.17). The force constant of the spring is k and the equilibrium length is L. Find (a) the kinetic energy of the system when the mass has a speed v and (b) the period of oscillation. (*Hint*: Assume that all portions of the spring oscillate in phase and that the velocity of a segment dx is proportional to the distance from the fixed end; that is, $v_x = (x/L)v$. Also, note that the mass of a segment of the spring is $dm = (m/L)dx$.

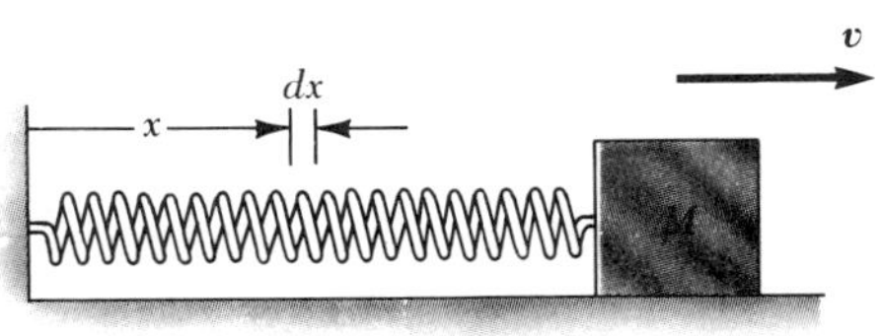

Figure P13.17

18. A 50-g mass attached to a spring moves on a horizontal, frictionless surface in simple harmonic motion. The amplitude is 16 cm, and the period is 4.0 s. At $t = 0$, the mass is released from rest at $x = 16$ cm, as in Figure 13.5 of the text. Find (a) the displacement as a function of time and its value at $t = 0.50$ s, (b) the magnitude and direction of the force acting on the mass at $t = 0.50$ s, (c) the minimum time required for the mass to reach the position $x = 8.0$ cm, (d) the velocity at any time t and the speed at $x = 8.0$ cm, and (e) the total mechanical energy and force constant of the spring.

Chapter 14

1. Two identical, isolated 2.00-kg particles are separated by 30.0 cm. What is the magnitude of the gravitational force exerted by one particle on the other?

2. Three 5.00-kg masses are located at the corners of an equilateral triangle having sides 0.250 m in length. Determine the magnitude and direction of the resultant gravitational force on one of the masses due to the other two.

3. Two stars of masses M and $4M$ are separated by a distance d. Determine the location of a point measured from M at which the net force on a third mass is zero.

4. Given that the Moon's period about the Earth is 27.32 days and the Earth-Moon distance is 3.84×10^8 m, estimate the mass of the Earth. Assume the orbit is circular. Why do you suppose your estimate is high?

5. A satellite of Mars has a period of 459 min. The mass of Mars is 6.42×10^{23} kg. From this information, determine the radius of the satellite's orbit.

6. At its aphelion, Mercury is 6.99×10^{10} m from the Sun, and at its perihelion, it is 4.60×10^{10} m from the Sun. If its orbital speed is 3.88×10^4 m/s at the aphelion, what is its orbital speed at the perihelion?

7. An isolated system consists of three 5.00-g particles located at the corners of an equilateral triangle with sides of 30.0 cm.
(a) Calculate the potential energy of the system. (b) If the particles are released simultaneously, where will they collide?

Chapter 14

8. The escape speed from the surface of the Earth is 11.2 km/s. Estimate the escape speed for a spacecraft from the surface of the Moon. The Moon has a mass $\frac{1}{81}$ that of Earth and a radius $\frac{1}{4}$ that of Earth.

9. (a) Calculate the minimum energy required to send a 3000-kg spacecraft from the Earth to a distant point in space where Earth's gravity is negligible. (b) If the journey is to take three weeks, what average power must the engines supply?

10. A *nonuniform* rod of length L is placed along the x axis at a distance h from the origin, as in Figure P14.10. The mass per unit length, λ, varies according to the expression $\lambda = \lambda_0 + Ax^2$, where λ_0 and A are constants. Find the force on a particle of mass m placed at the origin. (*Hint:* An element of the rod has a mass $dM = \lambda\, dx$.)

Figure P14.10

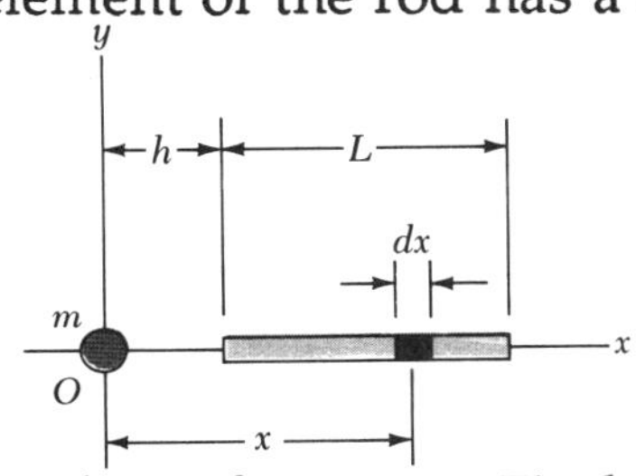

11. An 80.0-kg spherical shell has a radius of 0.500 m. Find the gravitational force exerted on a 50.0-g particle placed (a) 0.300 m from the center of the shell and (b) outside the shell 1.00 m from its center.

12. While approaching a planet circling a distant star, a space traveler determines the planet's radius to be half that of the Earth. After landing on the surface, the traveler finds the acceleration due to gravity to be twice that on the Earth's surface. Find the mass of the planet M_p in terms of the mass of the Earth M_E

Chapter 14

13. Four objects are located at the corners of a rectangle, as in Figure P14.13. Determine the magnitude and direction of the resultant force acting on the 2.00-kg mass at the origin.

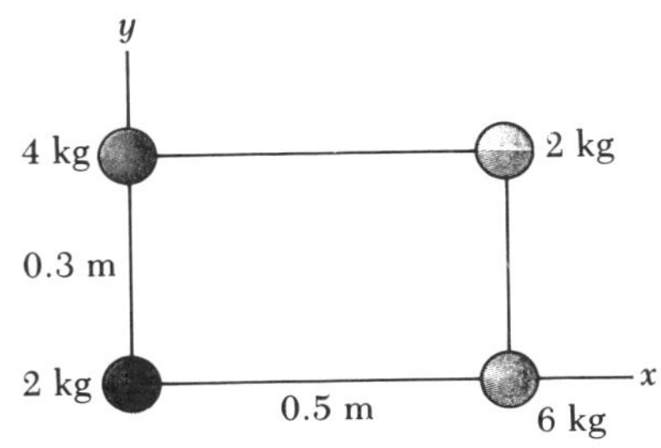

Figure P14.13

14. Two ocean liners, each having a mass of 40 000 metric tons, are moving on parallel courses 100 m apart. What is the magnitude of the acceleration of one of the liners toward the other due to the mutual gravitational attraction?

15. A satellite moves in an elliptical trajectory about the Earth as in Figure P14.15. The minimum and maximum distances from the surface of the Earth are 400 km and 3000 km. Find the speeds of the satellite when it is at these two distances.

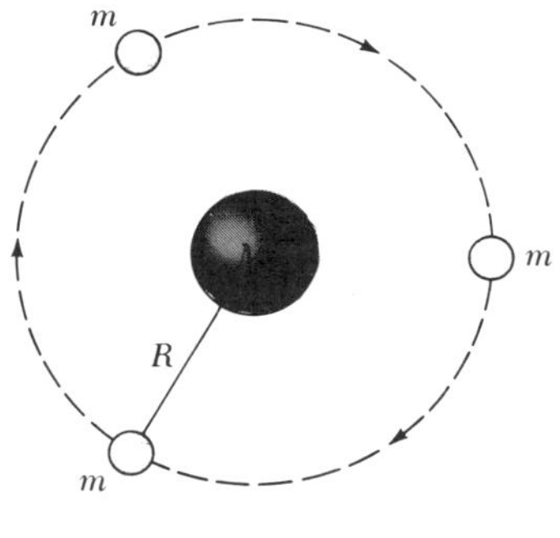

Figure P14.15

Chapter 15

1. At what depth in a lake is the absolute pressure equal to three times atmospheric pressure?

2. A magician is immersed in 4.0 m of water in a sealed trunk. If the lid of the trunk measures 0.70 m × 2.0 m, what is the force of the water on the trunk lid?

3. What is the fractional change in the density of sea water between the surface (where the pressure is equal to 1.0 atm) and a depth of 4.96 km (where the pressure is 500 atm)?

4. The U-shaped tube in Figure P15.4 contains mercury. What is the absolute pressure, P, on the left side if h = 20 cm? What is the gauge pressure?

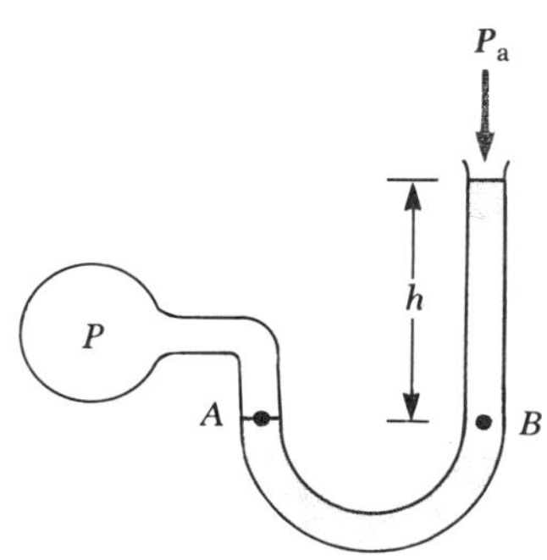

Figure P15.4

5. If the fluid in the barometer illustrated in Figure P15.5 is water, what will be the height of the water column in the vertical tube at atmospheric pressure?

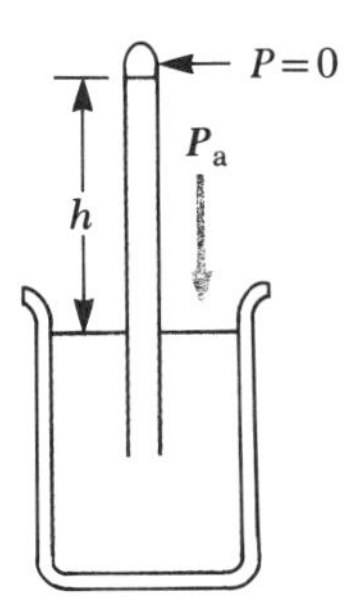

Figure P15.5

Chapter 15

6. Calculate the buoyant force on a solid object made of copper and having a volume of 0.2 m^3 if it is submerged in water. What is the result if the object is made of steel?

7. Show that only 11% of the total volume of an iceberg is above the water level. (Sea water has a density of 1.03×10^3 kg/m^3, and ice has a density of 0.92×10^3 kg/m^3.)

8. In air, an object weighs 15 N. When immersed in water, the same object weighs 12 N. When immersed in another liquid, it weighs 13 N. Find (a) the density of the object and (b) the density of the other liquid.

9. A rectangular air mattress has a length of 2.0 m, a width of 0.50 m, and a thickness of 0.08 m. If the mass of the material is 2.3 kg, what mass can just be supported by the mattress in water?

10. Water squirts from a syringe at a speed of 30 m/s. The diameter of the opening is 0.010 cm. Determine the pressure difference between the water inside the syringe and the air outside. Assume that the flow speed inside is zero. Express your result in pascals and atmospheres.

11. Each wing of an airplane has an area of 25 m^2. If the speed of the air is 50 m/s over the lower wing surface and 65 m/s over the upper wing surface, determine the weight of the airplane. (Assume the plane travels in level flight at constant speed at an elevation where the density of air is 1.0 kg/m^3. Also assume that all of the lift is provided by the wings.)

12. A 50.0-kg girl sits on a 1.00 m $\times$ 1.00 m $\times$ 0.0600 m raft made of solid Styrofoam. If the raft just supports the girl (that is, the raft is totally submerged), determine the density of the Styrofoam.

Chapter 15

13. A sample of copper is to be subjected to a hydrostatic pressure that will increase its density by 0.1%. What pressure is required?

14. One side of the U-shaped tube in Figure P15.14 is filled with a liquid of density ρ_1 while the other side contains a liquid of density ρ_2. If the liquids do not mix, show that $\rho_2 = (h_1/h_2)\,\rho_1$.

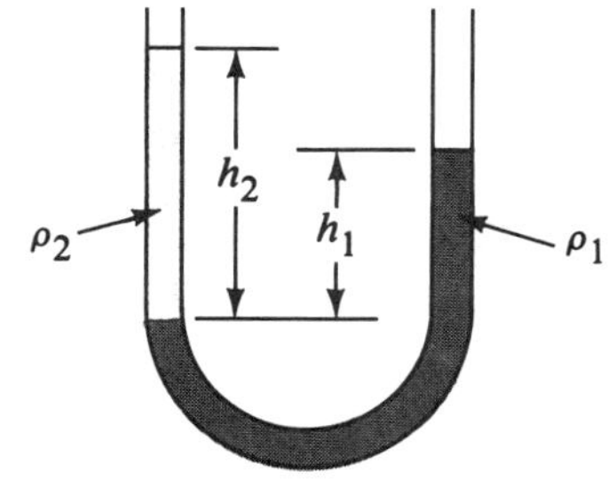

Figure P15.14

15. What power is required to pump 50 L/s of water through a horizontal pipe with a pressure difference of 1.0 atm?

16. Water flows through a horizontal pipe that gradually narrows so that the final inside diameter is one-half the original diameter. In the wide section the flow speed is 11.4 m/s and the pressure is 10.2 atm. Determine (a) the flow speed in the narrow section and (b) the pressure in the narrow section.

17. As a first approximation, the Earth's continents may be thought of as granite blocks floating in a denser rock (called peridotite) in the same way that ice floats in water. (a) Show that a formula describing this phenomenon is $\rho_g t = \rho_p d$, where ρ_g is the density of granite (2800 kg/m^3), ρ_p is the density of peridotite (3300 kg/m^3), t is the thickness of a continent, and d is the depth to which a continent floats in the peridotite. (b) If a continent rises 5 km above the surface of the peridotite (this surface may be thought of as the ocean floor), what is the thickness of the continent?

Chapter 15

18. Consider a composite "raft" consisting of two square slabs, each of side s, attached face to face. One slab has density ρ_1 and thickness h_1, while the other has density $\rho_2 > \rho_1$ and thickness h_2. (a) Find the average density $\bar{\rho}$ of the raft. (b) Assume that $\bar{\rho} < \rho_w$, so that the raft floats in water. The raft is placed in water with the denser slab on the bottom. Find d, the depth of the bottom surface of the raft. (c) If the raft is placed in water with the denser slab on the *top* find d', the depth of the bottom surface of the raft. Comment on your answer. (d) For which of the orientations described in part (b) and (c) is the gravitational potential energy of the entire system (consisting of the raft and the body of water in which it is floating) greater? Find the potential energy difference.

19. A cable of mass density ρ_c and diameter d extends vertically downward a distance h through water, and a block of mass M_b and density ρ_b is hung from the bottom end of the cable. Both ρ_c and ρ_b exceed ρ_w, the density of water. Find (a) the tension T_ℓ at the lower end of the cable, (b) the tension T_u at the upper end of the cable, and (c) the tensions T'_ℓ and T'_u that would exist at the lower and upper ends of the cable if the entire assembly were in air rather than water. (Neglect the buoyant force provided by the air.) (d) Evaluate T_ℓ, T_u, T'_ℓ, and T'_u for the case of a 100-meter steel cable supporting a prefabricated concrete object of mass 2.00 metric tons: $\rho_c = 7.86 \times 10^3$ kg/m^3, $d = 2 \times 10^{-2}$ m, $h = 100$ m, $M_b = 2.00 \times 10^3$ kg, and $\rho_b = 2.38 \times 10^3$ kg/m^3.

Chapter 16

1. A traveling wave pulse moving to the right along the x axis is represented by the wave function

$$y(x, t) = \frac{4}{2+(x-4t)^2}$$

where x and y are measured in centimeters and t is in seconds. Plot the shape of the waveform at $t = 0$, 1, and 2 s.

2. A traveling wave pulse moving to the left is described by the function

$$y = \frac{10}{5+(x+2t)^4}$$

where x and y are in centimeters and t is in seconds. (a) Compute the speed of this wave. (b) Plot this wave at two times and show that it is traveling to the left.

3. Two waves are described by $y_1 = 1/(ax - bt)$ and $y_2 = 1/(ax + bt)$, where a and b are constants. Find a relationship for the superposition of the two waves $y_1 + y_2$. What points along this wave are physically unrealistic and why?

4. The tension in a cord 15 m in length is 20 N. The measured transverse wave speed in the cord is 60 m/s. Calculate the total mass of the cord.

5. A rope 8.0 m long has a total mass of 500 g. The rope is stretched with a tension of 430 N. Find the speed of transverse waves on this rope.

Chapter 16

6. Transverse pulses travel with a speed of 200 m/s along a taut copper wire that has a diameter of 1.5 mm. What is the tension in the wire? (The density of copper is 8.92×10^3 kg/m^3.)

7. A transverse wave moving along a string in the positive x direction with a speed of 200 m/s has an amplitude of 0.700 mm and a wavelength of 20.0 cm. Determine (in SI units) the values of A, k, and ω in the equation describing the wave: $y = A\sin(kx - \omega t)$.

8. Determine the quantities (a) through (f) of Problem 24 when the wave train is described by $y = (0.20\text{ m})\sin 4\pi(0.40x + t)$. Again x and y are in meters and t is in seconds.

9. In Example 16.3 the sinusoidal wave was found to be described by $y = (15.0\text{ cm})\cos(0.157x - 50.3t)$, where x and y are in centimeters and t is in seconds. (a) Plot y versus x at $t = 0$ and $t = 0.0500$ s. (b) Determine the wave speed from this plot and compare your result with the value found in Example 16.3.

10. (a) Write the expression for y as a function of x and t for a sinusoidal wave traveling along a rope in the positive x direction with the following characteristics: $y_{max} = 8$ cm, $\lambda = 80$ cm, $f = 3$ Hz, and $y(0, t) = y_{max}$ at $t = 0$. (b) Determine the speed and angular wave number for the wave described in part (a).

11. A wire of mass 0.24 kg is 48 m long and under a tension of 60 N. An electric vibrator operating at an angular frequency of 80π rad/s is generating sinusoidal waves in the wire. The vibrator can supply energy to the wire at a maximum rate of 400 J/s. What is the maximum amplitude of the wave pulses?

Chapter 16

12. A 6.0-m segment of a long string contains four complete waves and has a mass of 180 g. The string is vibrating sinusoidally with a frequency of 50 Hz and a peak to peak displacement of 15 cm. (Peak to peak means the vertical distance from the farthest positive displacement to the farthest negative displacement.) (a) Write the function that describes this wave traveling in the positive x direction, and (b) determine the power being supplied to the string.

13. In Section 16.9 it is verified that $y_1 = A\sin(kx - \omega t)$ is a solution to the wave equation. The wave function $y_2 = B\cos(kx - \omega t)$ describes a wave that is shifted by 90° relative to y_1.
(a) Determine whether or not $y = y_1 + y_2$ is a solution to the wave equation. (b) Determine if $y = y_1 y_2$ is a solution to the wave equation.

14. What is the average rate at which energy is transmitted along the string in Problem 45 if μ = 75 g/m? What is the total energy per wavelength in the wave?

15. Consider the sinusoidal wave of Example 16.3, for which it was determined that $y = (15.0 \text{ cm})\cos(0.157x - 50.3t)$. At a given instant, let point A be at the origin and point B be the first point along x that is 60° out of phase with point A. What is the coordinate of point B?

16. Ocean waves with a wavelength of 120 m are traveling into a harbor at a rate of eight waves per minute. What is their speed?

Chapter 16

17. Two transverse wave pulses traveling in opposite directions along the x axis are represented by the wave functions

$$y_1(x, t) = \frac{6}{(x - 3t)^2}$$

$$y_2(x, t) = -\frac{3}{(x + 3t)^2}$$

where x and y are measured in centimeters and t is in seconds. Write a program which to obtain the shape of the composite wave $y_1 + y_2$ as a function of time. Use your program to plot the wave at $t = 0, 0.50, 1.0, 1.5, 2.0, 2.5$, and 3.0 s.

Chapter 17

1. The ocean floor is underlaid by a layer of basalt that constitutes the crust, or uppermost layer of the Earth in this region. Below this crust is found more dense peridotite rock, which forms the earth's mantle. The boundary between these two layers is called the Mohorovicic discontinuity ("Moho" for short). If an explosive charge is set off at the surface of the basalt, it generates a seismic wave that is reflected back at the Moho. If the velocity of this wave in basalt is 6.5 km/s, and the two-way travel time is 1.85 s, what is the thickness of this oceanic crust?

2. (a) What are the SI units of bulk modulus as expressed in Equation 17.2? (b) Show that the SI units of $\sqrt{B/\rho}$ are m/s, as required by Equation 17.1.

3. The density of acetone is 792 kg/m^3. Its bulk modulus has a temperature dependence given by the relation $B = -6 \times 10^6 T + 9.02 \times 10^8$ where B is in N/m^2 and T is in °C. What is the speed of sound in acetone at 20°C?

4. The pressure amplitude corresponding to the threshold of hearing is 2.9×10^{-5} Pa. At what frequency will a sound wave in air have this pressure amplitude if the displacement amplitude is 2.8×10^{-10} m?

5. Consider the sound wave whose displacement is described in Problem 15. What is the pressure variation at $x = 0$ when $t = \pi / 2\omega$? Take $\rho = 900$ kg/m^3.

6. (a) Calculate the intensity in W/m^2 of the wave described in Problem 14. (b) Express this intensity in dB.

Chapter 17

7. Calculate the pressure amplitude corresponding to a sound level of 120 dB (a rock concert).

8. The amplitude of the vibrations of a certain sound wave is increased by 50 percent. What is the corresponding increase in the decibel level of the sound?

9. At what speed should a supersonic aircraft fly so that the conical wavefront will have an apex half-angle of 40°?

10. A commuter train passes a passenger platform at a constant speed of 40 m/s. The train horn is sounded at its characteristic frequency of 320 Hz. (a) What change in frequency is observed by a person on the platform as the train passes? (b) What wavelength does a person on the platform observe as the train approaches?

11. A band is playing on a moving truck. The band strikes the note middle C (262 Hz), but it is heard by spectators ahead of the truck as C sharp [C# (277 Hz)]. How fast is the truck moving? Use 340 m/s for the speed of sound in air.

12. A painfully loud 2000-Hz tone has an intensity of 1.00 W/m^2. Calculate the corresponding air molecule amplitude of oscillation for an air density of 1.21 kg/m^3 and a sound speed of 343 m/s.

13. (a) The sound level of a jackhammer is measured as 130 dB and that of a siren as 120 dB. Find the ratio of the intensities of the two sound sources. (b) Two sources have measured intensities of $I_1 = 100\ \mu W/m^2$ and $I_2 = 200\ \mu W/m^2$. By how many dB is source 1 lower than source 2?

Chapter 17

14. The measured speed of sound in copper is 3560 m/s, and the density of copper is 8890 kg/m^3. Based on this information, by what percent would you expect a block of copper to decrease in volume when subjected to a uniform external (gauge) pressure of 2 atm?

15. The gas filling the tube shown in Figure 17.2 of the text is air at 20°C and at a pressure of 1.5×10^5 N/m^2. The piston shown is driven at a frequency of 600 Hz. The diameter of the piston is 10 cm, and the amplitude of its motion is 0.1 cm. What power must be supplied to maintain the oscillation of the piston?

16. Consider plane harmonic sound waves propagating in three different media at 0°C: air, water, and iron. Each wave has the same intensity (I_0) and the same angular frequency (ω_0).
(a) Compare the values of λ (the wavelength) in the three media.
(b) Compare the values of s_m (the displacement amplitude) in the three media. (*Hint:* Refer to Tables 15.1 and 17.1.) (c) Compare the values of ΔP_m (the pressure amplitude) in the three media.
(d) For $\omega_0 = 2000\pi$ rad/s and $I_0 = 10^{-6}$ W/m^2 (60 dB), evaluate λ, s_m, and ΔP_m for each of the three media.

Chapter 18

1. A harmonic wave is described by $y_1 = (8.0 \text{ m})\sin[2\pi(0.10x - 80t)]$, where y_1 and x are in meters and t is in seconds. Write an expression for a wave that has the same frequency, amplitude, and wavelength as y_1, but when added to y_1 will give a resultant with an amplitude of $8\sqrt{3}$ m.

2. Two speakers are arranged similarly to those shown in Figure 18.3. The distance between the two speakers is 2.00 m, and they are driven at a frequency of 1500 Hz. An observer is initially at a point 6.00 m along the perpendicular bisector of the line joining the two speakers. (a) What distance must the observer move along a line parallel to the line joining the two speakers before reaching the first minimum in intensity? (Use v = 343 m/s.) (b) At what distance from the perpendicular bisector will the observer find the first relative maximum in intensity?

3. Two identical sound sources are located along the y axis. Source S_1 is located at (0, 0.1) m and source S_2 is located at (0, −0.1) m. The two sources radiate isotropically at a frequency of 1715 Hz and the amplitude of each wave separately is assumed to be A. A listener is located along the y axis a distance of 5 m from source S_1. (a) What is the phase difference between the sound waves at the position of the listener? (b) What is the amplitude of the resultant wave at the location of the listener? (Use v = 343 m/s.)

4. Two waves which set up a standing wave in a long string are given by

$$y_1 = A\sin(kx - \omega t + \phi) \qquad y_2 = A\sin(kx + \omega t)$$

Show that the addition of the arbitrary phase angle (a) will change only the positions of the nodes, and (b) that the distance between nodes remains constant.

Chapter 18

5. A string with $L = 16$ m and m = 0.015 g/cm is stretched with a tension of 557 N ($\approx$ 125 lb). What is the highest harmonic of this string that is within the typical human's audible range (up to 20 000 Hz)?

6. Two pieces of steel wire having identical cross sections have lengths of L and $2L$. The wires are each fixed at both ends and stretched such that the tension in the longer wire is four times greater than that in the shorter wire. If the fundamental frequency in the shorter wire is 60 Hz, what is the frequency of the second harmonic in the longer wire?

7. A string of length L, mass per unit length m, and tension F is vibrating at its fundamental frequency. What effect will the following have on the fundamental frequency? (a) The length of the string is doubled with all other factors held constant. (b) The mass per unit length is doubled with all other factors held constant. (c) The tension is doubled with all other factors held constant.

8. A resonance condition is set up in a pipe with a tuning fork whose frequency is f. Write an expression for the length of the pipe that will cause it to resonate in its nth mode if the pipe is (a) open at both ends and (b) closed at one end. (Assume that the speed of sound is v.)

9. A glass tube (open at both ends) of length L is positioned near an audio speaker of frequency $f = 0.68$ kHz. For what values of L will the tube resonate with the speaker?

10. Determine the frequency corresponding to the first three resonances that produces standing waves in a 30.0-cm pipe when it is (a) open at both ends and (b) closed at one end.

Chapter 18

11. At a particular instant, the tube in Figure P18.11a is adjusted so that L, the length above the water surface, is 40 cm. The tuning fork in the figure is replaced by a variable-frequency oscillator that has a frequency range between 20 and 2000 Hz. What are the (a) lowest and (b) highest frequencies within this range that will excite resonant modes in the air column?

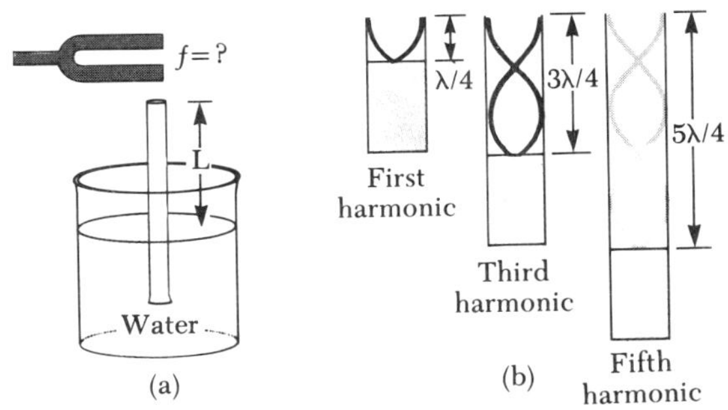

Figure P18.11

12. The range of a certain pipe organ is from 8 Hz to 30 000 Hz. What range of pipe lengths is necessary if they are (a) open at both ends and (b) closed at one end? (Assume that the speed of sound is 343 m/s.)

13. Two speakers are arranged as shown in Figure 18.3 of the text. For this problem, assume that point O is 12 m along the center line and the speakers are separated by a distance of 1.5 m. As the listener moves toward point P from point O, a series of alternating minima and maxima is encountered. The distance between the first minimum and the next maximum is 0.4 m. Using 344 m/s as the speed of sound in air, determine the frequency of the speakers. (Use the approximation $\sin\theta \approx \tan\theta$.)

14. Two pipes are each open at one end and are of adjustable length. Each has a fundamental frequency of 480 Hz at 300 K. The air temperature is increased in one pipe to 305 K. (a) If the two pipes are sounded together, what beat frequency will result? (b) By what percent should the length of the 305-K pipe be decreased to again match the frequencies? (Use $v = 331(T/273)^{1/2}$ m/s as the speed of sound in air, where T is the air temperature in K.)

Chapter 18

15. A speaker at the front of a room and an identical speaker at the rear of the room are being driven by the same oscillator at 456 Hz. A student walks at a uniform rate of 1.5 m/s along the length of the room. How many beats does the student hear per second?

16. To maintain a length of string under tension in a horizontal position, one end of the string is connected to a vibrating blade and the other end is passed over a pulley and attached to a mass. The mass of the string is 10 g, and its *total* length is 1.25 m. (a) When the suspended mass is 10 kg, the string vibrates in three equal length segments. Determine the vibration frequency of the blade. (Assume that the point where the string passes over the pulley and the point where it is attached to the blade are both nodes. Also, ignore the contribution to the tension due to the string's mass.) (b) What mass should be attached to the string if it is to vibrate in four equal segments?

Chapter 19

1. The gas thermometer in Figure P19.1 reads a pressure of 0.530 atm at the triple-point temperature of water. What pressure will it read at (a) the boiling point of water, and (b) the melting point of gold?

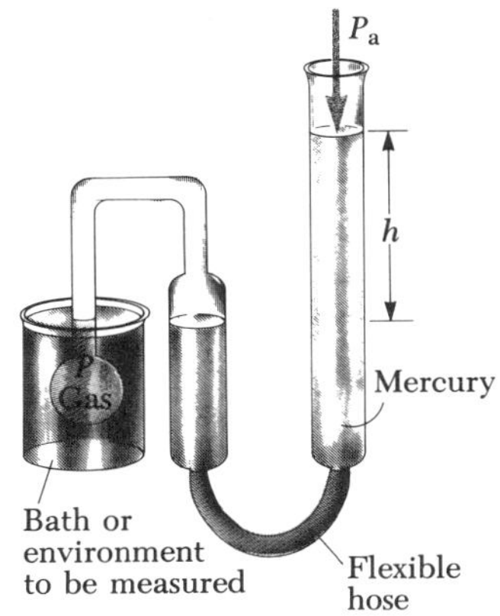

Figure P19.1

2. The pressure in a constant volume gas thermometer is 0.700 atm at 100.0°C and 0.512 atm at 0.0°C. (a) What is the temperature when the pressure is 0.0400 atm? (b) What is the pressure at 450.0°C?

3. The temperature of one northeastern state varies from 105°F in the summer to –25°F in winter. Express this range of temperature in degrees Celsius.

4. The boiling point of sulfur is 444.60°C. The melting point is 586.1°F below the boiling point. (a) Determine the melting point in degrees Celsius. (b) Find the melting and boiling points in degrees Fahrenheit.

5. A body cools from 350°C to –80°C. Express the change in temperature in (a) kelvin, and (b) Fahrenheit degrees.

Chapter 19

6. The length of the (former) standard meter is exactly 1 m at 0°C. This standard, kept at the International Bureau of Weights and Measures near Paris, France, is made of a platinum-iridium alloy that has an average coefficient of linear thermal expansion of 8.9×10^{-6} $(°C)^{-1}$. Assuming that all the numerical values are exact, and that α remains constant, find the length of the bar at 37°C.

7. A metal rod made of some alloy is to be used as a thermometer. At 0.0°C its length is 40.000 cm, and at 100.0°C its length is 40.060 cm. (a) What is the average linear expansion coefficient of the alloy? (b) What is the temperature when its length is 39.975 cm?

8. Calculate the fractional change in the volume ($\Delta V / V$) of an aluminum bar that undergoes a change in temperature of 30.0 °C.

9. A sample of ideal gas with a molecular mass of 4.0 g/mol is in a thermodynamic state having P = 1.2 atm, V = 8.8 L, and T = 85°C. Determine the mass of this sample of gas.

10. One mole of oxygen gas is at 6.00 atm and 27.0°C. (a) If the gas is heated at constant volume until the pressure triples, what is the final temperature? (b) If the gas is heated such that both the pressure and volume double, what is the final temperature?

11. A cylindrical diving bell 3.0 m in diameter and 4.0 m tall with an open bottom is submerged to a depth of 220 m in the ocean. The surface temperature is 25°C and the temperature 220 m down is 5°C. The density of sea water is 1025 kg/m^3. How high will the sea water rise in the bell when it is submerged?

Chapter 19

12. (a) Show that the average volume coefficient of thermal expansion for an ideal gas at constant pressure is given by $\beta = 1/T$, where T is the Kelvin temperature. Start with the definition of β and use the equation of state, $PV = nRT$.
(b) What value does this expression predict for β at 0°C? Compare this with the experimental values for helium and air in Table 19.2.

13. A swimmer has 0.82 L of dry air in his lungs when he dives into a lake. Assuming the pressure of the dry air is 95% of the external pressure at all times, what is the volume of the dry air at a depth of 10.0 m?

14. At 0.0°C, a container is completely full of ethyl alcohol. If the container does not expand when heated (and is pushed out a negligible extent when the ethanol begins to exert an outward pressure), what is the internal pressure when the temperature is 20.0°C? Express your answer in pascals and in atmospheres. (Refer to Table 19.2. The bulk modulus of ethyl alcohol is 1.0×10^9 N/m^2.)

Chapter 19

15. The measurement of the average coefficient of volume expansion for liquids is complicated by the fact that the container also changes size with temperature. Figure P19.15 shows a simple means for overcoming this difficulty. With this apparatus, $\overline{\beta}$ for a liquid is determined by measurements of the column heights, h_0 and h_t, of the liquid columns in the U-tube. One arm of the tube is maintained at 0°C in a water-ice bath, and the other arm is maintained at the temperature T_C in a thermal bath. The connecting tube is horizontal. Derive the expression for $\overline{\beta}$ in terms of h_0 and h_t.

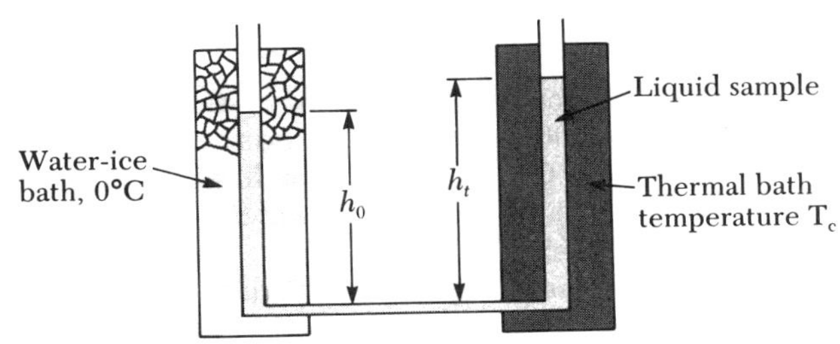

Figure P19.15

Chapter 20

1. A 50-g piece of cadmium is at 20°C. If 400 cal of heat is added to the cadmium, what is its final temperature?

2. Copper pellets, each of mass 1.0 g, are heated to 100°C. How many pellets must be added to 500 g of water initially at 20°C to make the final equilibrium temperature 25°C? (Neglect the heat capacity of the container.)

3. An aluminum calorimeter of mass 100 g contains 250 g of water. The calorimeter and vessel are in thermal equilibrium at 10°C. Two metallic blocks are placed in the water. One is a 50-g piece of copper at 80°C. The other sample has a mass of 70 g and is originally at 100°C. The entire system stabilizes at a final temperature of 20°C. (a) Determine the specific heat of the unknown sample. (b) Identfy the unknown metal using Table 20.1.

4. A Styrofoam cup contains 200 g of mercury at 0°C. To this is added 50 g of ethyl alcohol at 50°C and 100 g of water at 100°C. (a) What is the final temperature of the mixture? (b) How much heat was gained or lost by the mercury, the alcohol, and the water? (The specific heat of mercury is 0.033 cal/g·°C, that of ethyl alcohol, 0.58 cal/g·°C. The heat capacity of the Styrofoam cup is negligible.)

5. A 50 g copper calorimeter contains 250 g of water at 20°C. How much steam must be condensed into the water to make the final temperature of the system 50°C?

6. A beaker of water sits in the Sun until it reaches an equilibrium temperature of 30°C. The beaker is made of 100 g of aluminum and contains 180 g of water. In an attempt to cool this system down, 100 g of ice at 0°C is added to the water. (a) If the final equilibrium temperature is 0°C, determine how much ice remains. (b) Repeat this for the case when 50 g of ice is used.

7. Using the fact that 1 atm = 1.013×10^5 N/m^2, verify the conversion 1 L·atm = 101.3 J = 24.2 cal.

8. (a) Determine the work done by a fluid that expands from i to f in Figure P20.8. (b) How much work is performed by the fluid if it is compressed from f to i along the same path?

Figure P20.8

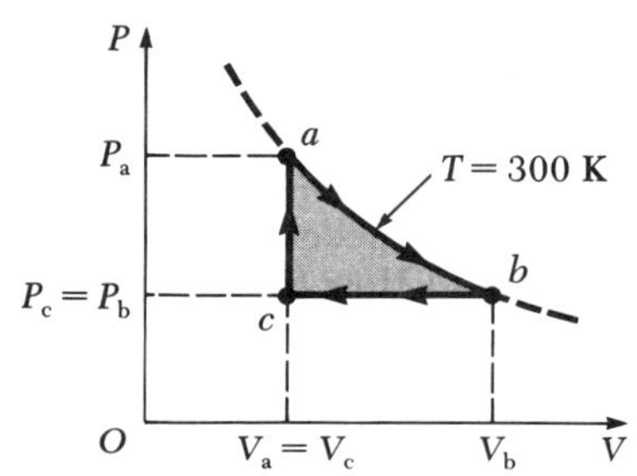

9. A gas expands from I to F as in Figure P20.9. The thermal energy added to the gas is 400 J when the gas goes from I to F along the diagonal path. (a) What is the change in internal energy of the gas? (b) How much thermal energy must be added to the gas for the indirect path IAF to give the same change in internal energy?

Figure P20.9

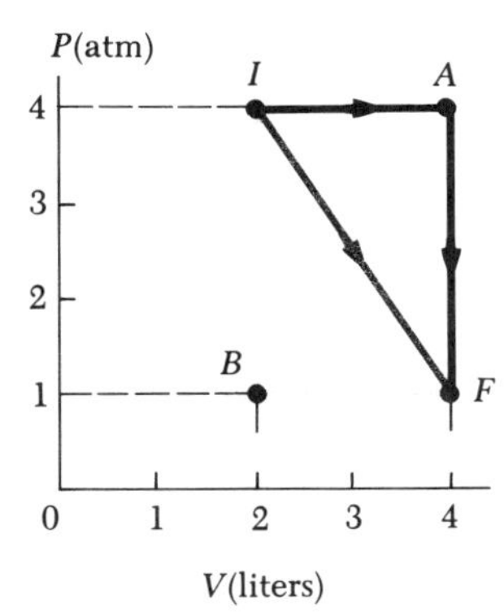

10. Consider the cyclic process described by Figure P20.10. If Q is negative for the process BC and ΔU is negative for the process CA, determine the signs of Q, W, and ΔU associated with each process.

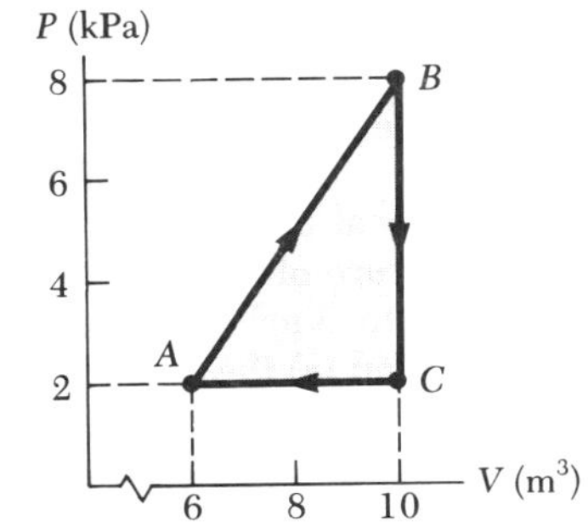

Figure P20.10

11. One mole of gas initially at a pressure of 2.0 atm and a volume of 0.30 L has an internal energy equal to 91 J. In its final state, the pressure is 1.5 atm, the volume is 0.80 L, and the internal energy equals 182 J. For the three paths IAF, IBF, and IF in Figure P20.11, calculate (a) the work done by the gas and (b) the net thermal energy transferred in the process.

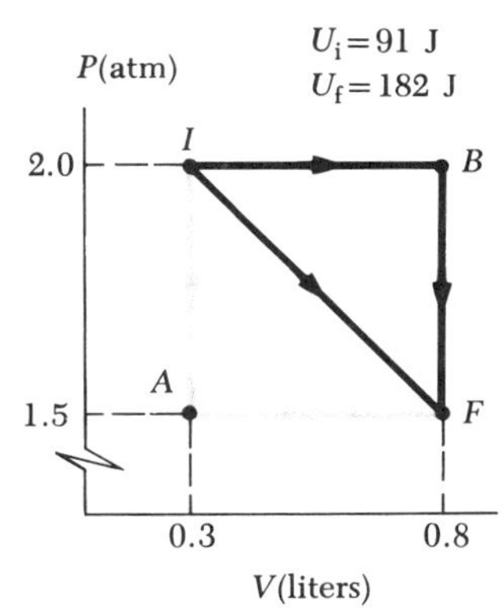

Figure P20.11

12. Nitrogen gas (m = 1.00 kg) is confined in a cylinder with a movable piston exposed to normal atmospheric pressure. A quantity of heat (Q = 25 000 cal) is added to the gas in an isobaric process, and its internal energy increases by 8000 cal. (a) How much work is done by the gas? (b) What is the change in volume?

Chapter 20

13. One mole of argon is confined in a cylinder with a movable piston at 1 atm and at 300 K. The gas is heated slowly and isobarically to 400 K. The measured value of the molar specific heat at constant pressure for argon in this temperature range is $C_p = 2.5043R$, and the measured value of PV/nT is $0.99967R$. Calculate in units of R, carried to two decimals, the following quantities: (a) the work done by the expanding gas; (b) the amount of heat added to the gas; (c) the increase in the internal energy of the gas.

14. An iron rod 20.0 cm long with a diameter of 1.00 cm has one end immersed in an ice water bath while the other end is in a steam tank at 100.0°C. Assume enough time has elapsed such that a uniform thermal gradient has been established along the rod. (a) Determine the rate of heat flow along the rod. (b) How fast is the ice is melting at the cold end? (c) How fast is steam being condensed at the hot end to maintain this uniform gradient? (d) What is the thermal gradient along the rod?

15. The Earth's thermal gradient, as measured at the surface, is 30 °C/km, and the Earth's radius is 6400 km. Assume that this gradient remains the same all the way to the center of the Earth. What is the temperature of the Earth at its center if we take the surface temperature to be 0°C? Do you think this is a reasonable answer or is it necessary to refine our assumption on how the thermal gradient changes with depth?

16. The density of water is 999.17 kg/m^3 at 14.5°C and 999.02 kg/m^3 at 15.5°C. (a) Calculate the work done against the surrounding atmosphere when 2.00 kg of water expands as its temperature is increased from 14.5°C to 15.5°C. (b) Compare this work with the required heat input of 8370 J.

Chapter 20

17. One mole of an ideal gas is contained in a cylinder with a movable piston. The initial pressure, temperature, and volume are P_0, V_0, and T_0. Find the work done by the gas for the following processes and show each process on a PV diagram: (a) an isobaric compression to $V_0/2$, (b) an isothermal compression to $4P_0$, (c) an isovolumetric process to $3P_0$.

18. A gas expands from 2.0 m^3 to 6.0 m^3 along two paths as described in Figure P20.18. The heat added to the gas along the path *IAF* is 4.0×10^5 cal. Find (a) the work done by the gas along the path *IAF*, (b) the work done along the path *IF*, (c) the change in internal energy of the gas, and (d) the heat transferred in the process along the path *IF*.

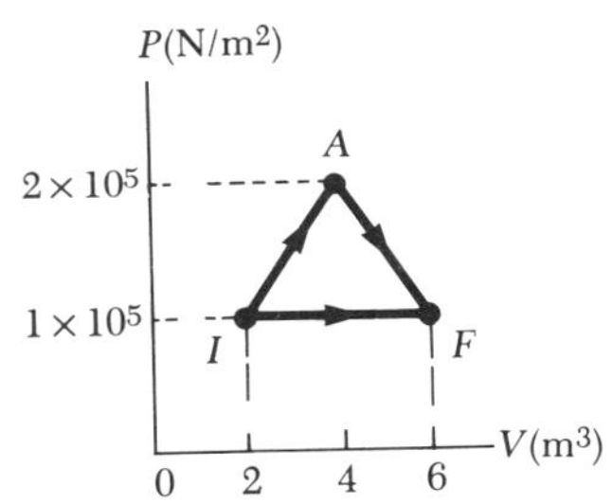

Figure P20.18

19. An aluminum kettle has a circular cross section and is 8.00 cm in radius and 0.30 cm thick. It is placed on a hot plate and filled with 1.0 kg of water. If the bottom of the kettle is maintained at 101.0°C and the inside at 100.0°C, find (a) the rate of heat flow into the water and (b) the time it takes for all of the water to boil away. (Neglect heat transferred from the sides.)

20. Using the data in Example 20.5 and Table 20.2, calculate the change in internal energy when 2.00 cm^3 of liquid helium gas at 4.2 K is converted to helium gas at 273.15 K and atmospheric pressure. (Assume that the molar heat capacity of helium gas is 24.9 J/mol·K, and note that 1.00 cm^3 of liquid helium is equivalent to 3.10×10^{-2} mol.)

Chapter 21

1. Two moles of oxygen gas is heated from 300 K to 320 K. How much heat is transferred to the gas if the process occurs at (a) constant volume and (b) constant pressure?

2. The heat capacity, C, of a monatomic gas measured at constant pressure is 62.3 J/K. Find (a) the number of moles of gas, (b) the heat capacity at constant volume, and (c) the internal energy of the gas at 350 K.

3. Consider 3 mol of an ideal gas. (a) If the gas is monatomic, find the total heat capacity at constant volume and at constant pressure. (b) Repeat part (a) for a diatomic gas in which the molecules rotate but do not vibrate.

4. How much internal energy is contained in 1.00 m^3 of air at 0.0°C?

5. The specific heat at constant volume for neon gas is $C_v = 0.149$ kcal/kg·K. Calculate the mass of a neon atom.

6. Show that Equation 21.20 follows from Equation 21.19 for an adiabatic process. (*Note:* $PV = nRT$ applies during the process.)

7. An ideal gas ($\gamma = 1.40$) expands slowly and adiabatically. If the final temperature is one third the initial temperature, (a) by what factor does the volume change? (b) By what factor does the pressure change?

8. One mole of an ideal monatomic gas ($\gamma = 1.67$) initially at 300 K and 1.00 atm is compressed slowly and adiabatically to one fourth its initial volume. Find its final pressure and temperature.

9. Examine Table 21.2 and explain why SO_2 has a higher value of C_v than the other polyatomic gases at 300 K.

Chapter 21

10. Fifteen identical particles have the following speeds: one has speed 2.0 m/s; two have speed 3.0 m/s; three have speed 5.0 m/s; four have speed 7.0 m/s; three have speed 9.0 m/s; two have speed 12.0 m/s. Find (a) the average speed, (b) the rms speed, and (c) the most probable speed of these particles.

11. Calculate the most probable speed, average speed, and rms speed for nitrogen gas molecules at 900 K. Compare your results with the values obtained from Figure 21.12 in the textbook.

12. Use Figure 21.12 to estimate the number of nitrogen molecules with speeds between 400 m/s and 600 m/s at (a) 300 K and (b) 900 K.

13. Using the data in Figure 21.12, estimate the fraction of N_2 molecules that have speeds in the range 1000 m/s to 1200 m/s at 900 K. The total number of molecules is 1.00×10^5.

14. Consider the situation described in Problem 36. Show that, in general, the elapsed time interval between arrival of the sound signal through the pipe and through the surrounding air is

$$\Delta t = \frac{\ell(v_m - v_a)}{v_m v_a}$$

where v_m is the speed of sound in the metal, v_a is the speed of sound in air, and ℓ is the length of the pipe.

Chapter 22

1. In each cycle of its operation, a certain refrigerator absorbs 100 J from the cold reservoir and expels 130 J. (a) What is the power required to operate the refrigerator if it works at 60 cycles/s? (b) What is the coefficient of performance of the refrigerator?

2. Over one cycle a Carnot engine absorbs 24 MJ of thermal energy along an isotherm at 140°C and ejects thermal energy along a 14°C isotherm. Determine the amount of thermal energy ejected and the Carnot efficiency.

3. An electrical generating plant has a power output of 500 MW. The plant uses steam at 200°C and exhausts water at 40°C. If the system operates with one half the maximum (Carnot) efficiency, (a) at what rate is thermal energy expelled to the environment? (b) If the thermal energy goes into a river whose flow rate is 1.2×10^6 kg/s, what is the rise in temperature of the river?

4. A heat engine operating between 200°C and 80°C achieves 20% of the maximum possible efficiency. What energy input will enable the engine to perform 1.0×10^4 J of work?

5. An ideal gas is taken through a Carnot cycle. The isothermal expansion occurs at 250°C, and the isothermal compression takes place at 50°C. If the gas absorbs 1200 J of thermal energy during the isothermal expansion, find (a) the thermal energy expelled to the cold reservoir in each cycle and (b) the net work done by the gas in each cycle.

6. A gasoline engine using an ideal diatomic gas ($\gamma = 1.4$) operates between 300 K and 1500 K. Determine its compression ratio if it has an efficiency of 20%. Compare this efficiency to that of a Carnot engine operating between the same temperatures.

7. What is the coefficient of performance of a refrigerator that operates with Carnot efficiency between –3°C and 27°C?

8. What is the entropy decrease in 1 mol of helium gas cooled at 1 atm from room temperature (293 K) to 4 K? (C_p of helium = 21 J/mol·K).

9. When 1 mol of an ideal monatomic gas is heated slowly at constant volume from 300 K to 400 K, what is the change in its entropy?

10. If 3200 J of thermal energy flows from a heat reservoir at 500 K to another reservoir at 300 K through a conducting metal rod, find the change in entropy of (a) the hot reservoir, (b) the cold reservoir, (c) the metal rod, and (d) the Universe.

11. A 70-kg log falls from a height of 25 m into a lake. If the log, the lake, and the air are all at 300 K, find the change in entropy of the universe for this process.

12. A substance undergoes a Carnot cycle during which its temperature changes by a factor of 2 and its entropy changes by a factor of 4. (a) Sketch a graph of the cycle using temperature-entropy (*T*-*S*) axes. (b) What is the physical significance of the area enclosed by the graph?

13. If 200 g of water at 20°C is mixed with 300 g of water at 75°C, find (a) the final equilibrium temperature of the mixture and (b) the change in entropy of the system.

14. What is the minimum amount of work that must be done to extract 400 J of thermal energy from a massive object at 0°C while rejecting thermal energy to a 20°C reservoir?

Chapter 22

15. An athlete whose mass is 70 kg drinks 16 ounces (453.6 g) of refrigerated water. The water is at a temperature of 35°F. (a) Neglecting the temperature change of the body resulting from the water intake (so that the body is regarded as a reservoir at 98.6°F), find the entropy increase of the system. (b) Assume that the entire body is cooled by the drink and that the average specific heat of a human is equal to the specific heat of water. Neglecting any other heat transfers and any metabolic heat release, find the athlete's temperature after drinking the cold water, given an initial body temperature of 98.6°F. Under *these* assumptions, what is the entropy increase of the system? Compare your result with that of (a).

16. Suppose an ideal gas is used as the working substance in the Carnot cycle depicted in Figure P22.16. Remember that the internal energy of an ideal gas depends ontemperature only. (a) If the gas absorbs $Q_h = 3.0$ kJ during the isothermal expansion from *A* to *B*, how much work is performed by the gas? (b) If the gas performs 1.0 kJ of work during the adiabatic expansion from *B* to *C*, what is the change in its internal energy? (c) How much work must be performed on the gas from *C* to *D* in order for it to eject $Q_c = 2.0$ kJ of thermal energy? (d) How much work must be performed on the gas to take it from *D* to *A*? (e) What is the net amount of work performed by the gas during the cycle? (f) What is the thermodynamic efficiency of the cycle?

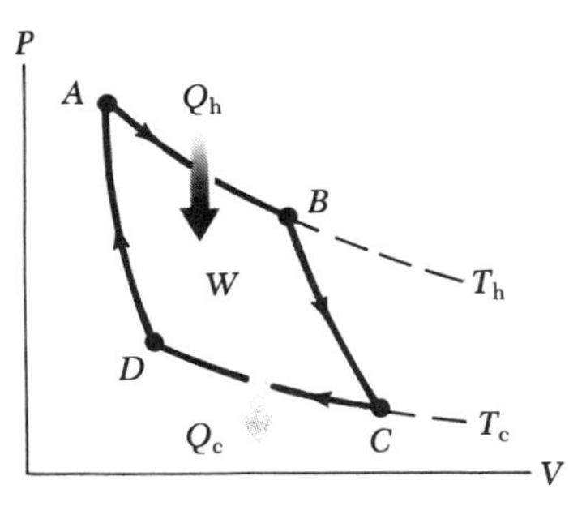

Figure P22.16

Chapter 22

17. During each cycle, a refrigerator extracts 360 kJ of energy from the low-temperature reservoir and ejects 480 kJ of energy to the high-temperature reservoir. Determine (a) the work done on the refrigerant and (b) the coefficient of performance.

Chapter 23

1. Two protons in a molecule are separated by 3.8×10^{-10} m. Find the electrostatic force exerted by one on the other.

2. A 6.7-μC charge is located 5.0 m from a –8.4-μC charge. Find the electrostatic force exerted by one on the other.

3. A 1.3-μC charge is located on the x axis at $x = -0.50$ m, a 3.2-μC charge is located on the x axis at $x = 1.5$ m, and a 2.5-μC charge is located at the origin. Find the net force on the 2.5-μC charge.

4. Two identical small metal spheres 1.19 m apart attract each other with a force of 0.0853 N. They are then brought into electrical contact with each other so that the net charge is shared equally. When returned to a separation of 1.19 m, the spheres repel each other with a force of 0.0196 N. Find the charge originally on each sphere.

5. Four point charges are situated at the corners of a square of sides a as in Figure P23.5. Find the resultant force on q.

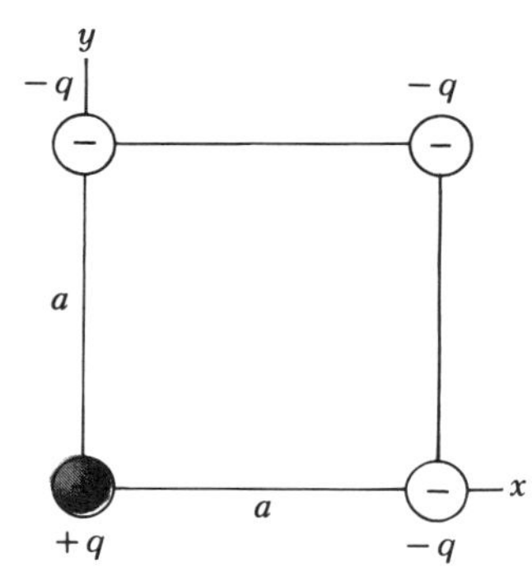

Fig. P23.5

6. Four identical point charges are located at the corners of a rigid square plate. Prove that the net torque about the center of the plate is zero.

Chapter 23

7. A point charge of $-5.2\ \mu C$ is located at the origin. Find the electric field (a) on the x axis at $x = 3$ m, (b) on the y axis at $y = -4$ m, (c) at the point with coordinates $x = 2$ m, $y = 2$ m.

8. Find the total electric field along the line of the two charges shown in Figure P23.8 at the point midway between them.

Fig. P23.8

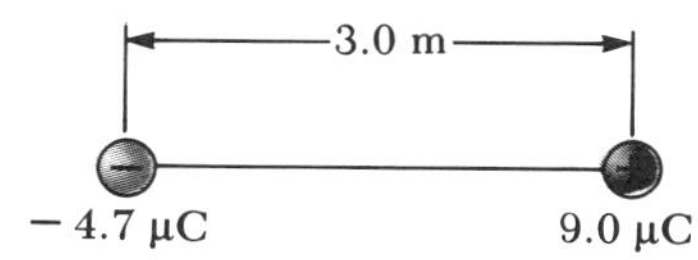

9. Four identical point charges ($q = +6\ \mu C$) are located on a rectangle as shown in Figure P23.9, with $L = 80$ cm and $W = 20$ cm. Calculate the resultant electric field at the center of the rectangle.

Figure P23.9

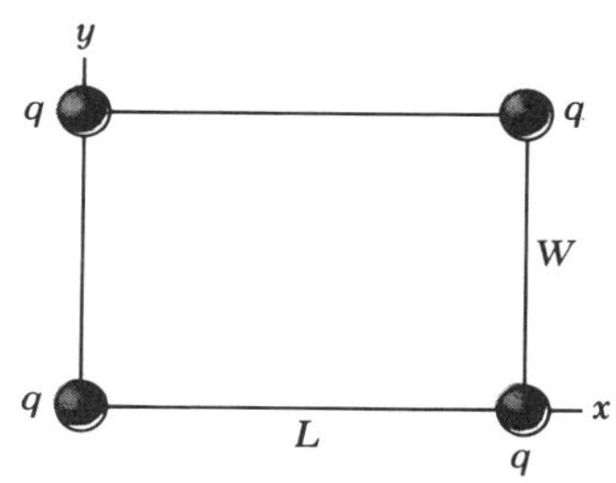

10. Five equal point charges $-q$ are placed symmetrically around a circle of radius R as in Figure P23.10. Calculate the electric field E at the center of the circle.

Figure P23.10

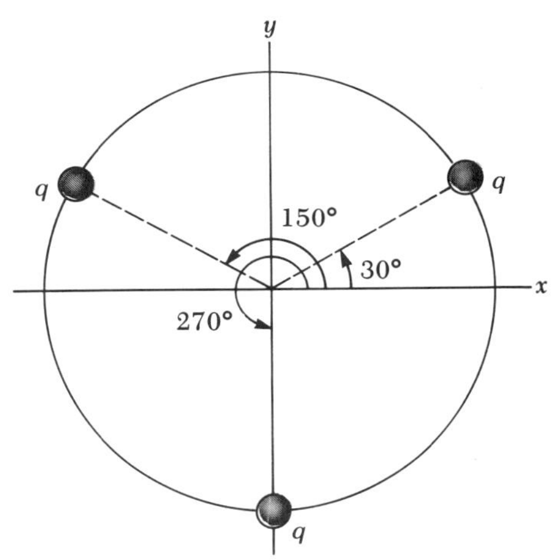

Chapter 23

11. Three identical point charges ($q = +2.7\ \mu C$) are placed on the corners of an equilateral triangle whose sides have a length of 35 cm (Fig. P 23.11). What is the magnitude of the resultant electric field at the center of the triangle?

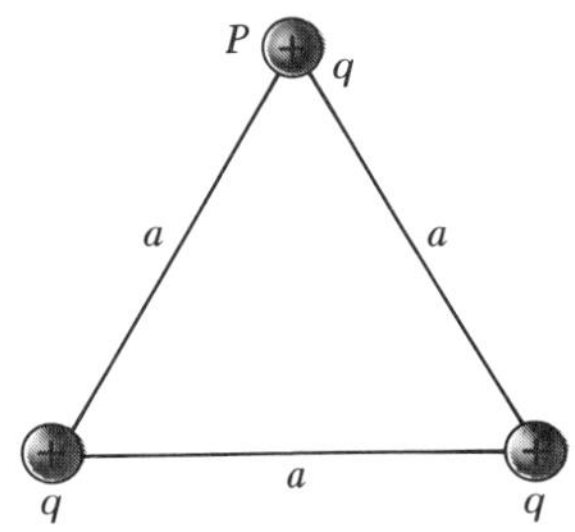

Figure P23.11

12. Four charges are at the corners of a square as in Figure P23.12. (a) Find the magnitude and direction of the electric field at the position of the charge $-q$, the coordinates of which are $x = a$, $y = a$. (b) What is the electric force on this charge?

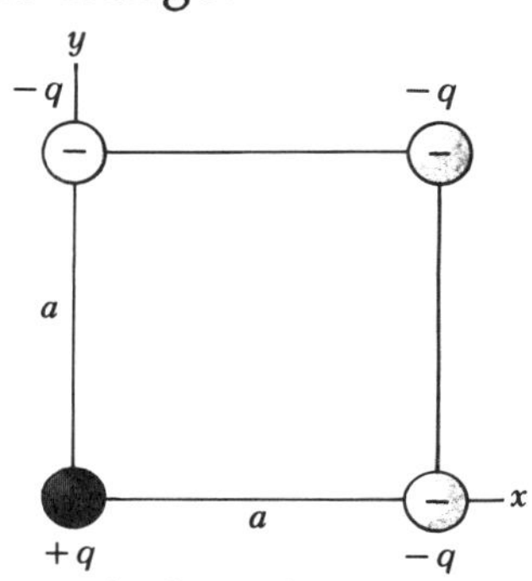

Figure P23.12

13. In Figure P23.13, determine the point (other than ∞) at which the total electric field is zero.

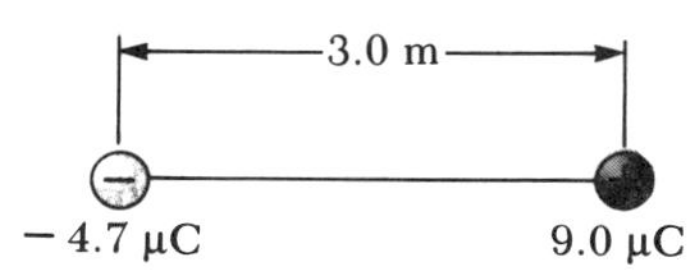

Figure P23.13

14. A sphere of radius 4.0 cm has a net charge of $+39\ \mu C$. (a) If this charge is uniformly distributed throughout the volume of the sphere, what is the volume charge density? (b) If this charge is uniformly distributed on the sphere's surface, what is the surface charge density?

Chapter 23

15. Four equal positive point charges are at the corners of a square. Sketch the electric field lines in the plane of the square.

16. A proton and an electron both start from rest and from the same point in a uniform electric field of 370 N/C. How far apart are they after 1.00 ms? (Ignore the attraction between them. If you like, you might imagine the experiment first with the proton only and then with the electron only.)

17. Two small spheres of charge Q are suspended from strings of length ℓ that are connected at a common point. One sphere has mass m; the other has mass $2m$. Assume the angles θ_1 and θ_2 that the strings make with the vertical are small. (a) How are θ_1 and θ_2 related? (b) Show that the distance r between the spheres is

$$r \cong \left(\frac{3k_e Q^2 \ell}{2mg} \right)^{1/3}$$

18. Two rings each of radius R have their axes oriented along the same line and are separated by a distance $2R$. One ring has uniform linear charge density $+\lambda$, and the other has uniform linear charge density $-\lambda$ Find (a) the electric field at point P_1 midway between the two rings on the common axis and (b) the electric field at point P_2 on the axis a distance R outside the negatively charged ring.

Chapter 23

19. Many molecules possess an electric dipole moment because the center of distribution of the positive charge (protons) does not exactly coincide with that of the negative charge (electrons). The electric dipole moment of a water molecule in its gaseous state is 6.24×10^{-30} C·m. (a) If a water molecule is placed in an electric field of 10^4 N/C, calculate the maximum torque that the field can exert on the molecule. (b) Find the range of the potential energies that the molecule may have in this field.

20. In the Millikan oil drop experiment, the droplets are so tiny that they appear only as points of light in the microscope used to observe them. In order to find the radius (and hence the mass) of each droplet, we allow them to fall freely under gravity. The retarding force **F** exerted by the viscous air on a sphere of radius r moving with speed v through air is given by Stokes' law, $F = 6\pi\eta r v$, where η is the coefficient of viscosity. (a) Find the SI units for η . (b) Show that when a falling droplet achieves a constant "terminal" velocity (signifying that the total force on the droplet is zero), the following relation is true, thus allowing the radius of the droplet to be determined:

$$v = \frac{2gr^2}{9\eta}(\rho_0 - \rho_a)$$

Chapter 24

1. A flat surface having an area of 3.2 m^2 is rotated in a uniform electric field of magnitude $E = 6.2 \times 10^5$ N/C. Calculate the electric flux through this area when the electric field is (a) perpendicular to the surface, (b) parallel to the surface, and (c) makes an angle of 75° with the surface.

2. A nonuniform electric field is given by the expression $\mathbf{E} = ay\,\mathbf{i} + bz\,\mathbf{j} + cx\,\mathbf{k}$, where a, b, and c are constants. Determine the electric flux through a rectangular surface in the xy plane, extending from $x = 0$ to $x = w$ and from $y = 0$ to $y = h$.

3. An electric field is given by $\mathbf{E} = az\,\mathbf{i} + bx\,\mathbf{k}$, where a and b are constants. Determine the electric flux through the shaded triangular surface in Figure P24.3.

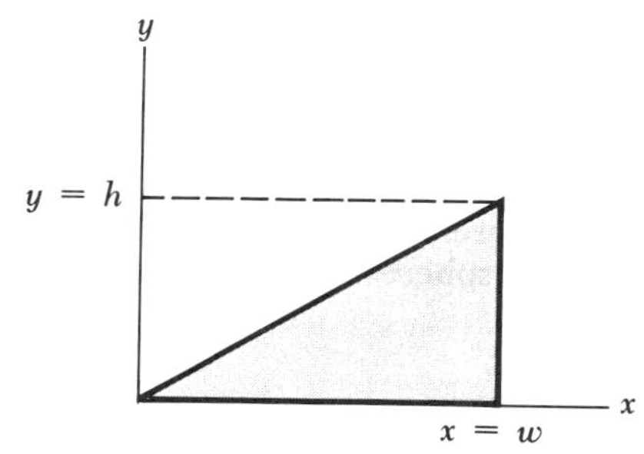

Fig. P24.3

4. A cone with a circular base of radius R stands upright so that its axis is vertical. A uniform electric field $\mathbf{E}$ is applied in the vertical direction. Show that the flux through the cone's surface (not counting its base) is given by $\pi R^2 E$.

5. A point charge of 5.0 μC is located at the center of a sphere having a radius of 12 cm. What is the electric flux through the surface of this sphere?

6. The electric field in the Earth's atmosphere is $E = 100$ N/C, pointing downward. Determine the electric charge on the Earth.

Chapter 24

7. Five charges are placed in a closed box. Each charge (except the first) has a magnitude which is twice that of the previous one placed in the box. If all charges have the same sign and if (after all charges have been placed in the box) the net electric flux through the box is 4.8×10^7 N·m^2/C, what is the magnitude of the smallest charge? Does the answer depend on the size of the box?

8. The electric field everywhere on the surface of a hollow sphere of radius 0.750 m is 8.90×10^2 N/C and points radially toward the center of the sphere. (a) What is the net charge inside the sphere? (b) What can you conclude about the nature and distribution of this charge?

9. A cube of sides 10 cm is centered at the origin. A point charge of 2.0 μC is located on the y axis at y = 20 cm. (a) Sketch the electric field lines for the point charge. (b) What is the net flux through the surface of the cube? (c) Repeat (a) and (b) if a second point charge of 4.0 μC is located at the center of the cube. (Neglect the lines that go through the edges and corners.)

10. An insulating sphere of radius 10.0 mm has a uniform charge density 6.00×10^{-3} C/m^3. Calculate the electric flux through a concentric spherical surface with radius (a) r = 5.00 mm, (b) r = 10.0 mm, and (c) r = 25.0 mm.

11. An insulating sphere of 10.0-cm radius has a uniform charge density throughout its volume. If the magnitude of the electric field 5.0 cm from the center is 8.6×10^4 N/C, what is the magnitude of the electric field 15 cm from the center?

Chapter 24

12. A cylindrical shell of radius 7.0 cm and length 240 cm has its charge uniformly distributed on its surface. The magnitude of the electric field at a point 19 cm radially outward from its axis (measured from the midpoint of the shell) is 3.6×10^4 N/C. Use approximate relationships to find (a) the net charge on the shell and (b) the electric field at a point 4.0 cm from the axis, measured from the midpoint.

13. A conducting spherical shell of radius 15 cm carries a net charge of −6.4 μC uniformly distributed on its surface. Find the electric field at points (a) just outside and
(b) inside the shell.

14. A square plate of copper of sides 50.0 cm is placed in an extended electric field of magnitude 8.00×10^4 N/C directed perpendicular to the plate. Find (a) the charge density of each face and (b) the total charge on each face.

15. A very large, thin, flat plate of aluminum of area A has a total charge Q uniformly distributed over its surfaces. If the same charge is spread uniformly over the upper surface of an otherwise identical glass plate, compare the electric fields just above the center of the upper surface of each plate.

16. A solid copper sphere 15.0 cm in radius has a total charge of 40.0 nC. Find the electric field (a) 12.0 cm, (b) 17.0 cm, and (c) 75.0 cm from the center of the sphere. (d) How would your answers change if the sphere were hollow?

17. Consider the data given in Problem 45. Calculate the net charge enclosed by a concentric spherical gaussian surface having (a) $r = 1.0$ cm, (b) $r = 3.0$ cm, (c) $r = 4.5$ cm, and (d) $r = 7.0$ cm.

Chapter 24

18. The electric field on the surface of an irregularly shaped conductor varies from 5.60×10^4 N/C to 2.80×10^4 N/C. Calculate the local surface charge density at the point on the surface where the radius of curvature of the surface is (a) greatest and (b) smallest.

19. The flux of any vector field **V** through a surface can be defined as

$$\Phi_v = \int_S \mathbf{V} \cdot d\mathbf{A}$$

Radial fields whose magnitudes vary inversely as the square of the distance, such as the gravitational field

$$\mathbf{g} = \frac{\mathbf{F}_g}{m}$$

also obey Gauss's Law. Using Gauss's (gravitational) law, find the gravitational field g at a point located a distance r from the center of the Earth where $r < R_e$. Pretend that the Earth's density is uniform.

Chapter 25

1. Concentric spherical surfaces surrounding a point charge at their center are equipotential surfaces. The intersections of these surfaces with a plane through their common center are equipotential lines. How much work is done in moving a charge q a distance s along an arc of an equipotential of circular shape and of radius R ?

2. Suppose an electron is released from rest in a uniform electric field whose strength is 5.9×10^3 V/m. (a) Through what potential difference does it pass after moving 1.0 cm? (b) How fast is the electron moving after it has traveled 1.0 cm?

3. A proton moving in a uniform electric field experiences an increase in kinetic energy of 5.0×10^{-18} J after being displaced 2.0 cm in a direction parallel to the field. What is the magnitude of the electric field?

4. A uniform electric field of magnitude 325 V/m is directed in the negative y direction in Figure P25.4. The coordinates of point A are (– 0.200, – 0.300) m, and those of point B are (0.400, 0.500) m. Calculate the potential difference $V_B - V_A$ using path ACB.

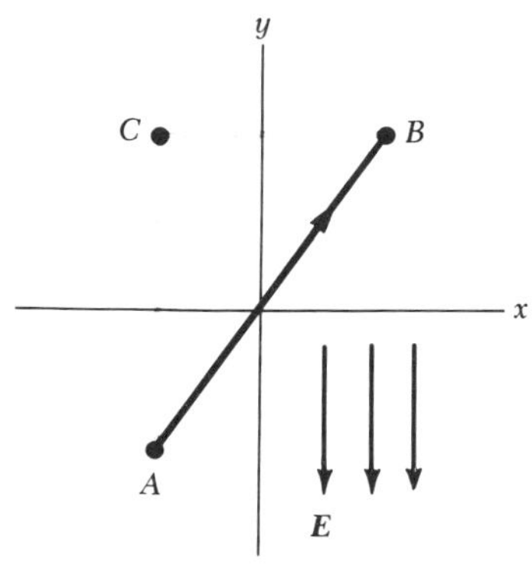

Fig. P25.4

Chapter 25

5. For the situation described in Problem 18, calculate the change in electric potential while going from A to B along the direct path. Which point is at the higher potential?

6. A uniform electric field of magnitude 250 V/m is directed in the positive x direction. (a) Suppose a + 12-μC charge moves from the origin to the point $(x, y) = (20 \text{ cm}, 50 \text{ cm})$. Through what potential difference does it move? (b) What is the change in its potential energy?

7. A + 2.8-μC charge is located on the y axis at $y = +1.6$ m and a – 4.6-μC charge is located at the origin. Calculate the net electric potential at the point (0.40, 0) m.

8. Calculate the value of the electric potential at point P due to the charge configuration shown in Figure P25.8. Use the values $q_1 = 5.0$ μC, $q_2 = -10$ μC, $a = 0.40$ m, and $b = 0.50$ m.

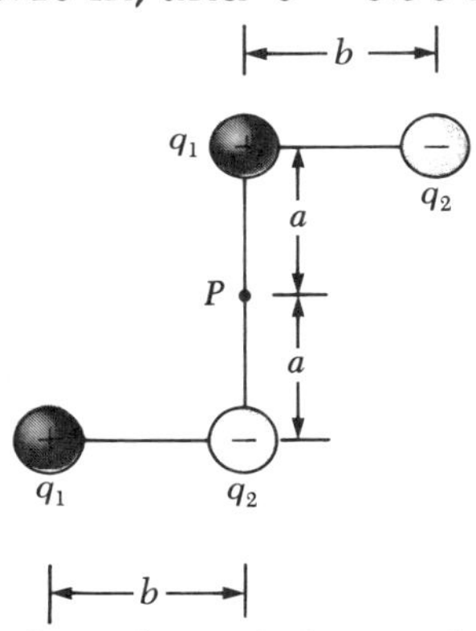

Fig. P25.8

9. A charge $q_1 = -9.0$ μC is located at the origin, and a second charge $q_2 = -1.0$ μC is located on the x axis at $x = 0.70$ m. Calculate the electric potential energy of the pair.

10. Four 5.00-μC point charges are located at the corners of a 30.0 cm by 40.0 cm rectangle. Calculate the electric potential energy stored in this charge configuration.

Chapter 25

11. Four charges are located at the corners of a rectangle as in Figure P25.11. How much energy is expended in removing the two 4.0-μC charges to infinity?

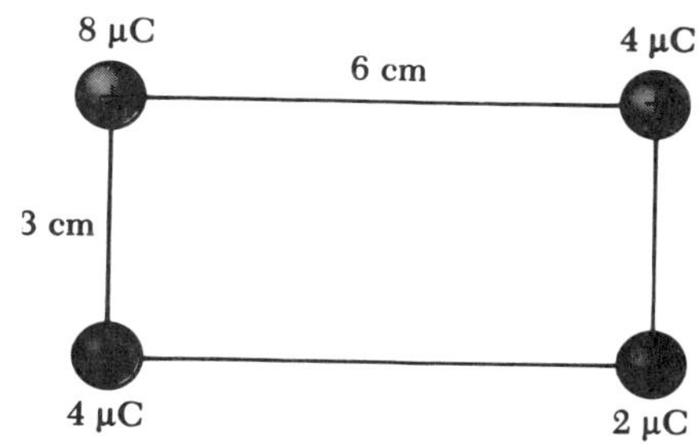

Fig. P25.11

12. The electric potential over a certain region of space is $V = 3x^2y - 4xz - 5y^2$ V. Find (a) the electric potential and (b) the components of the electric field at the point (1, 0, 2), where all distances are in meters.

13. Use the exact result from Example 25.4 to evaluate V at $x = 3a$ for the described dipole when $a = 2.00$ mm and $q = 3.00$ μC. Compare this answer to that obtained when you use the approximate formula valid when $x \gg a$.

14. Use the exact result for the dipole potential from Example 25.4 to find the electric field at any point where $x > a$ on the axis of the dipole. Evaluate E at $x = 3a$ if $a = 2.00$ mm and $q = 3.00$ μC.

15. Two spherical conductors of radii r_1 and r_2 are connected by a conducting wire as shown in Figure P25.15. If $r_1 = 0.94$ m, $r_2 = 0.47$ m, and the field at the surface of the smaller sphere is 890 N/C, calculate the total charge on the larger sphere assuming it is initially uncharged. Assume that the separation of the spheres is large compared to their radii.

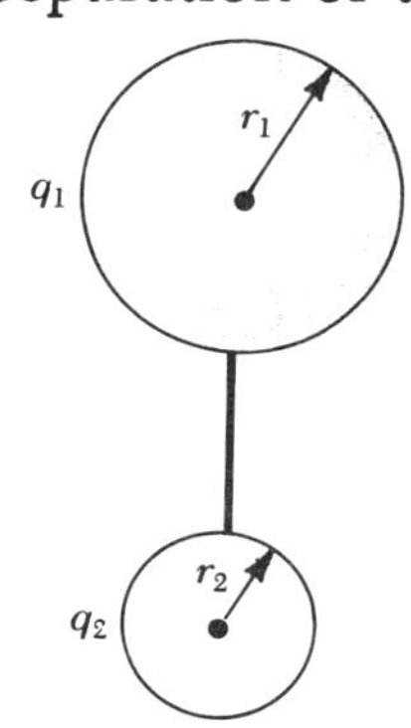

Figure P25.15

Chapter 25

16. What power in watts must a Van de Graaff generator deliver if it produces a 100-mA beam of protons at an energy of 12 MeV?

17. All corners except one of a 1.00-m cube are occupied by charges of 1.00 μC. What is the electric potential at the empty corner?

18. The electric potential just outside a charged conducting sphere is 200 V, and 10 cm farther from the center of the sphere the potential is 150 V. Find (a) the radius of the sphere and (b) the charge on it.

19. A Van de Graaff generator is operating so that the potential difference between the high-voltage electrode and the charging needles (points *B* and *A* in Figure P25.24) is 1.5×10^4 V. Calculate the power required to drive the belt (against electrical forces) at an instant when the effective current delivered to the high-voltage electrode is 500 mA.

20. A uniformly charged rod is located along the x axis as in Figure 23.14. The total charge on the rod is 16×10^{-10} C, $d = 1.0$ m, and $\ell = 2.0$ m. Estimate the electrical potential at $x = 0$ by approximating the rod to be (a) a point charge at $x = 2.0$ m, (b) two point charges (each of charge 8.0×10^{-10} C) at $x = 1.5$ m and $x = 2.5$ m, and (c) four point charges (each of charge 4.0×10^{-10} C) at $x = 1.25$ m, $x = 1.75$ m, $x = 2.25$ m, and $x = 2.75$ m. (d) Write a program that enables you to extend your calculations to 256 equally spaced point charges, and compare your result with that given by the exact expression

$$V = k_e \frac{Q}{\ell} \ln\left(\frac{\ell + d}{d}\right)$$

Chapter 25

21. A ring of radius 1.0 m has a uniform charge per unit length and a total charge of 16×10^{-10} C. The ring lies in the yz plane, and its center is at $x = 0$, as in Figure P25.21. Estimate the electric potential along the x axis at $x = 2.0$ m by approximating the ring to be (a) a point charge located at the origin, (b) two point charges (each of charge 8.0×10^{-10} C) diametrically opposite each other on the ring, and (c) four point charges (each of charge 4.0×10^{-10} C) symmetrically spaced on the ring. (d) Write a program that enables you to extend your calculations to 64 point charges equally spaced on the ring, and compare your result with that given by the exact expression, Equation 25.18.

Figure P25.21

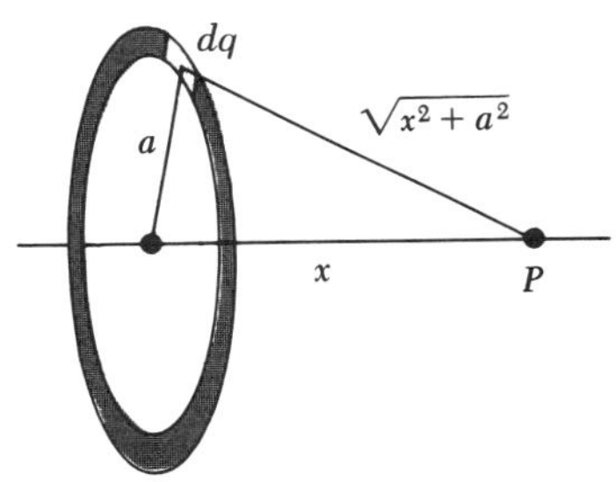

Chapter 26

1. Show that the units $C^2/N \cdot m$ equal 1 F.

2. Two parallel wires are suspended in a vacuum. When the potential difference between them is 52.0 V, each wire has a charge of 73.0 pC (the two charges are of opposite sign). Calculate the capacitance of the parallel-wire system.

3. An isolated conducting sphere can be considered as one element of a capacitor (the other element being a concentric sphere of infinite radius). (a) If the capacitance of this system is 9.1×10^{-11} F, what is the radius of the sphere? (b) If the potential at the surface of the sphere is 2.8×10^4 V, what is the corresponding surface charge density?

4. A circular parallel-plate capacitor with a spacing $d = 3.0$ mm is charged to produce an electric field strength of 3.0×10^6 V/m. What plate radius R is required if the stored charge Q is 1.0 μC?

5. One plate of a parallel-plate capacitor (with area $A = 100\ \text{cm}^2$) is grounded. The ungrounded plate carries a fixed charge Q. When the ungrounded plate is moved 0.50 cm farther away from the grounded plate, the potential difference between the plates increases by 200 V. Determine the magnitude of Q.

6. (a) Figure P26.6 shows a network of capacitors between the terminals *a* and *b*. Reduce this network to a single equivalent capacitor. (b) Determine the charge on the 4.0-µF and 8.0-µF capacitors when the capacitors are fully charged by a 12-V battery connected to the terminals. (c) Determine the potential difference across each capacitor.

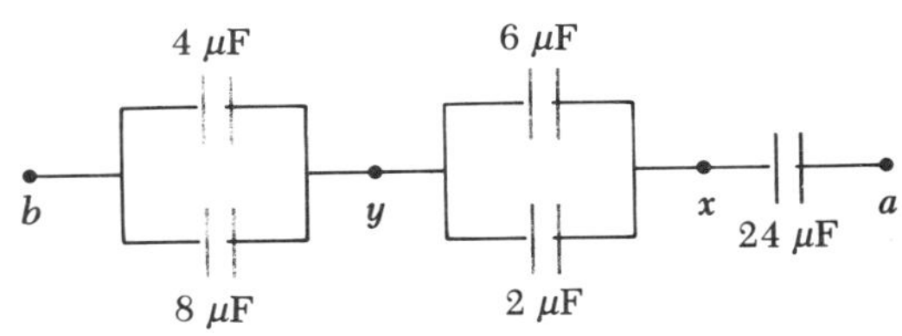

Fig. P26.6

7. Consider the group of capacitors shown in Figure P26.7.
(a) Find the equivalent capacitance between points *a* and *b*.
(b) Determine the charge on each capacitor when the potential difference between *a* and *b* is 12 V.

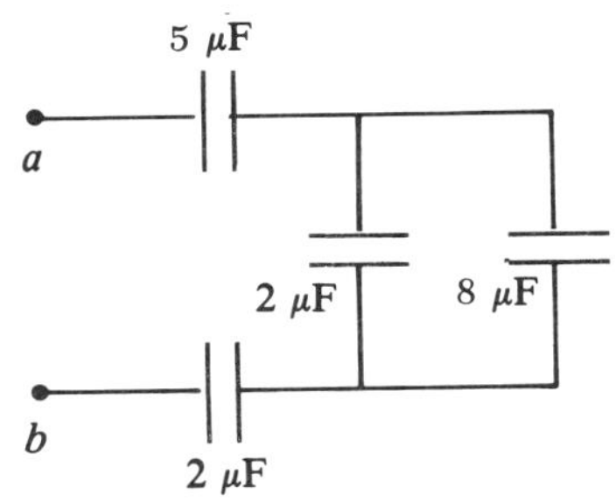

Fig. P26.7

8. Consider the combination of capacitors shown in Figure P26.8.
(a) What is the equivalent capacitance between points *a* and *b* ?
(b) Determine the charge on each capacitor if $V_{ab} = 4.8$ V.

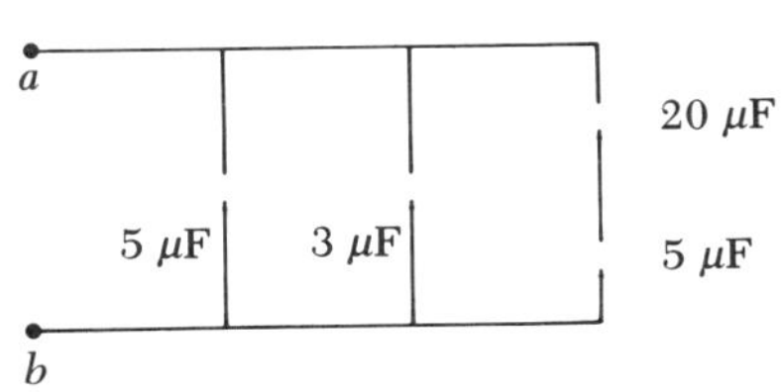

Fig. P26.8

Chapter 26

9. Find the equivalent capacitance between points *a* and *b* in the capacitor network shown in Figure P26.9.

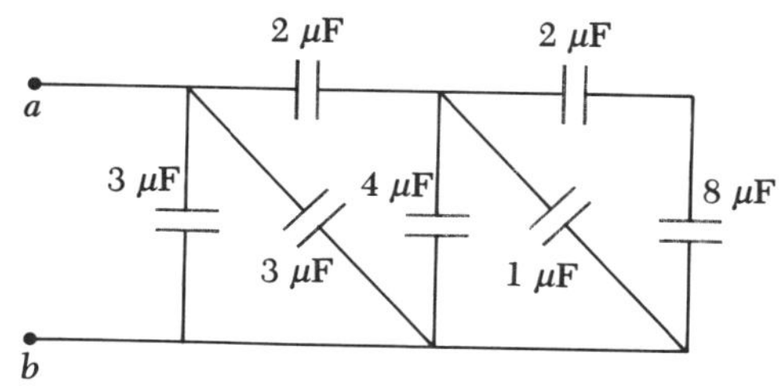

Fig. P26.9

10. What total energy is stored in the group of capacitors shown in Figure P26.10 if V_{ab} = 15 V?

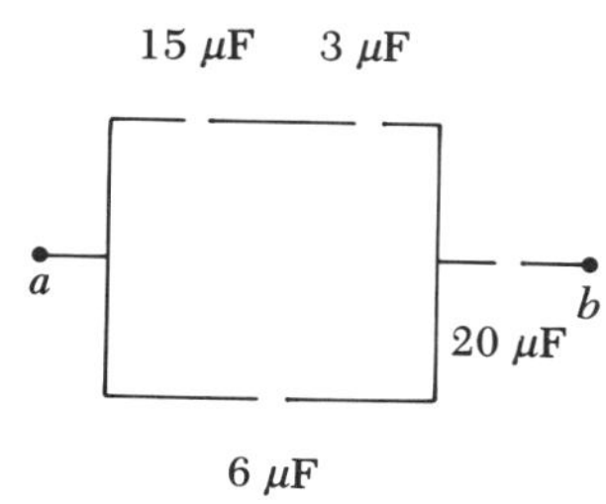

Figure P26.10

11. Two capacitors, C_1 = 25 μF and C_2 = 5.0 μF, are connected in parallel and charged with a 100-V power supply. (a) Calculate the total energy stored in the two capacitors. (b) What potential difference would be required across the same two capacitors connected in series in order that the combination store the same energy as in part (a)?

12. The energy density in a parallel-plate capacitor is 2.10×10^{-9} J/m^3. What is the magnitude of the electric field in the region between the plates?

13. Use Equation 26.14 to calculate the energy stored in the field of a simple spherical capacitor. Show that $U = Q^2/2C$.

Chapter 26

14. Determine (a) the capacitance and (b) the maximum voltage which can be applied to a Teflon-filled parallel-plate capacitor having a plate area of 175 cm^2 and insulation thickness of 0.0400 mm. (See Table 26.1 for other dielectric properties.)

15. A parallel-plate capacitor having air between its plates is charged to 31.5 V. The capacitor is then isolated from the charging source and the volume between the plates is filled with Plexiglas. Determine the new potential difference across the capacitor. The dielectric constant of Plexiglas is 3.12.

16. Let Q_0 be the greatest charge that can be placed on the plates of an air-filled parallel-plate capacitor without causing electrical breakdown. Let Q be the greatest charge that can be placed on the plates of this same capacitor when the gap between the plates is filled with neoprene rubber. What is the ratio Q/Q_0?

17. When the voltage applied to a capacitor increases from 80 V to 110 V, the charge on the capacitor increases by 9.0×10^{-5} C. Determine the capacitance.

18. A parallel-plate capacitor is to be constructed using Pyrex glass as a dielectric. If the capacitance of the device is to be 0.20 μF and it is to be operated at 6000 V, (a) calculate the minimum plate area required. (b) What is the energy stored in the capacitor at the operating voltage? (For Pyrex, use $\kappa = 5.6$. Its dielectric strength is 14×10^6 V/m.)

Chapter 27

1. Calculate the current in the case for which 3×10^{12} electrons pass a given cross section of a conductor each second.

2. The current I (in A) in a conductor depends on time as $I = 2t^2 - 3t + 7$, where t is in s. What quantity of charge moves across a section through the conductor during the interval $t = 2$ s to $t = 4$ s?

3. Calculate the number of free electrons per cubic meter for gold, assuming one free electron per atom.

4. A 2.4-m length of wire that is 0.031 cm^2 in cross section has a measured resistance of 0.24 Ω. Calculate the conductivity of the material.

5. What is the resistance of a tungsten filament 15 cm in length and 0.002 cm in diameter at 20°C?

6. What diameter copper wire has a resistance per unit length of 3.28×10^{-3} Ω/m at 20°C?

7. Determine the resistance at 20°C of a 1.5-m length of platinum wire that has a diameter of 0.10 mm.

8. What is the resistance of a device that operates with a current of 7 A when the applied voltage is 110 V?

9. Calculate the percentage change in the resistance of a carbon filament when it is heated from room temperature to 160°C.

10. An aluminum rod has a resistance of 1.234 Ω at 20°C. Calculate the resistance of the rod at 120°C by accounting for the change in both the resistivity and dimensions of the rod.

11. How much current is being supplied by a 200-V generator delivering 100 kW of power?

12. If a 55-Ω resistor is rated at 125 W (the maximum allowed power), what is the maximum allowed operating voltage?

13. An electric heater with a resistance of 20 Ω requires 100 V across its terminals. A built-in switching circuit repetitively turns the heater on for 1 s and off for 4 s. (a) How much energy is produced by the heater in 1 h? (b) What is the average power delivered by the heater over a period of one cycle?

14. An electric heater operating at full power draws a current of 8 A from a 110-V circuit. (a) What is the resistance of the heater? (b) Assuming the resistance R remains constant, at what voltage and current would the heater dissipate 750 W?

15. (a) A sheet of copper ($\rho = 1.7 \times 10^{-8}$ Ω·m) is 2 mm thick and has surface dimensions of 8 cm × 24 cm. If the long edges are joined to form a hollow tube 24 cm in length, what is the resistance between the ends? (b) What mass of copper would be required to manufacture a spool of copper cable 1500 m in length and having a total resistance of 4.5 Ω?

Chapter 28

1. A television repairman needs a 100-Ω resistor to repair a malfunctioning set. He is temporarily out of resistors of this value. All he has in his tool box are a 500-Ω resistor and two 250-Ω resistors. How can the desired resistance be obtained from the resistors on hand?

2. Find the equivalent resistance between points *a* and *b* in Figure P28.2.

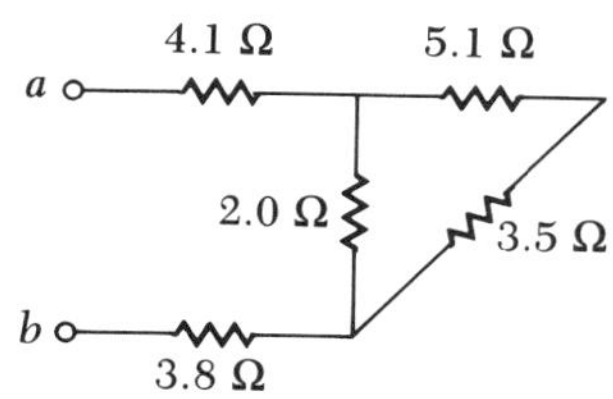

Fig. P28.2

3. Evaluate the effective resistance of the network of identical resistors shown in Figure P28.3.

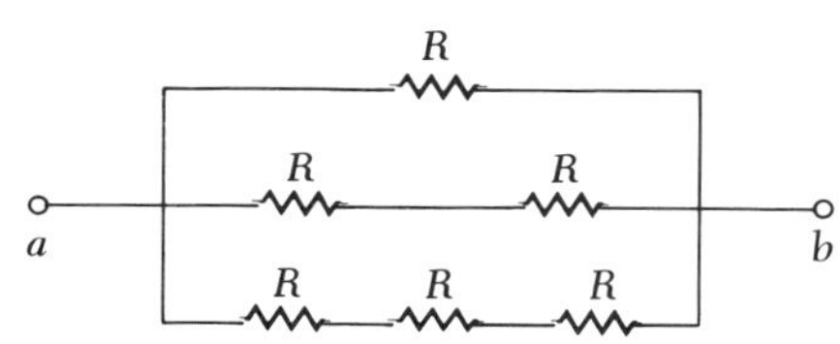

Fig. P28.3

4. In Figure P28.4, each resistor has a resistance of 1 Ω. Suppose that a given current *I* enters at *a* and comes out at *b*. Using the symmetry of the network, show that the equivalent resistance from *a* to *b* is $\frac{2}{3}$ Ω. (*Hint:* What would the resistance be if the vertical resistors were absent?)

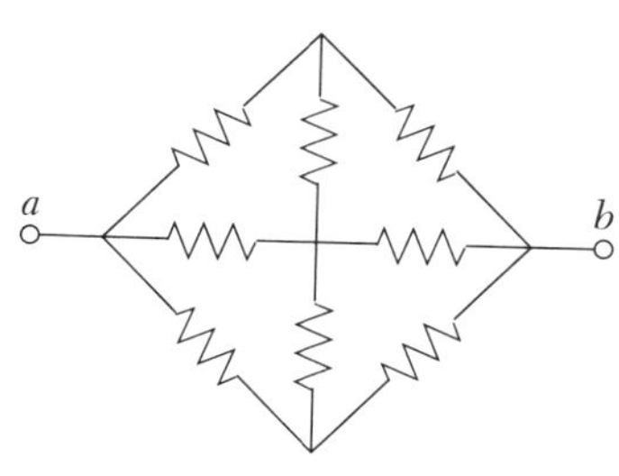

Fig. P28.4

Chapter 28

5. In Figure P28.5, the potential at point d equals zero. Determine the potentials at points (a) a, (b) b, and (c) c.

Fig. P28.5

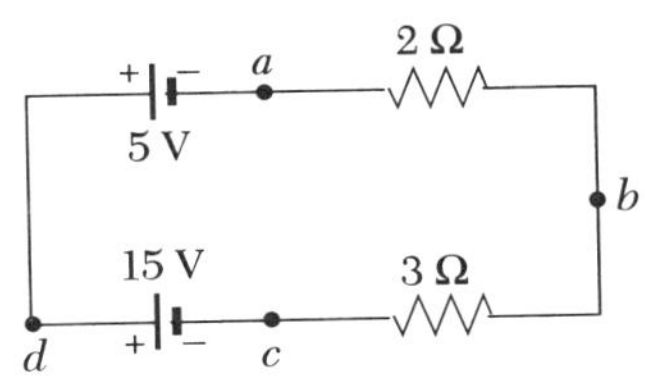

6. (a) Find the value of I_1 and I_3 in Figure P28.6 if the 4.0-V battery is replaced by a 5.0-μF capacitor. (b) Determine the charge on the capacitor.

Fig. P28.6

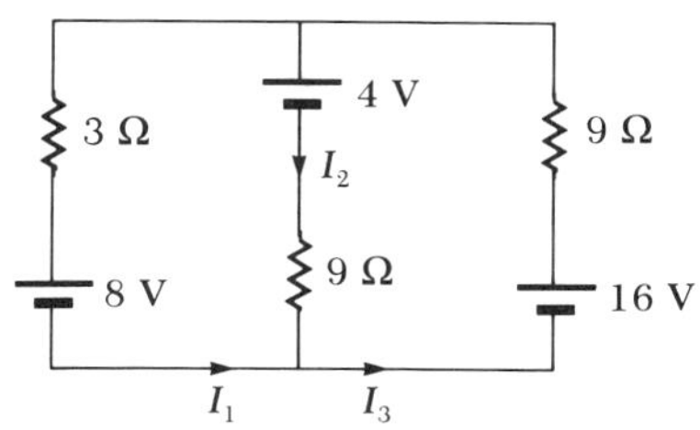

7. Using Kirchhoff's rules, find the current in each resistor in Figure P28.7.

Fig. P28.7

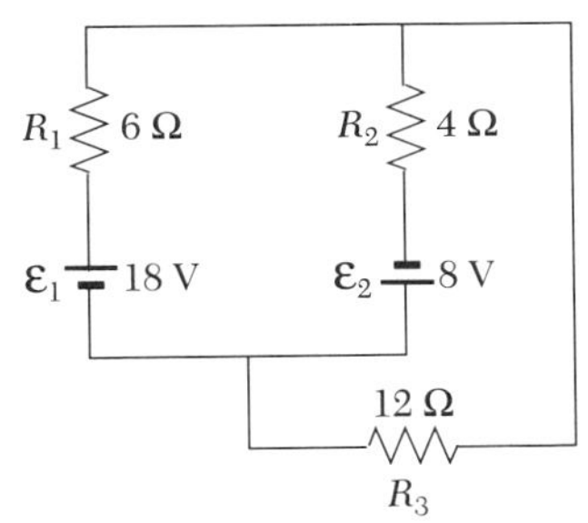

8. Consider the capacitor-resistor combination described in Problem 48. (a) How much energy is stored initially in the charged capacitor? (b) If the capacitor is completely discharged through the resistor, how much energy is dissipated as heat in the resistor?

Chapter 28

9. Consider a galvanometer with an internal resistance of 60.0 Ω. If the galvanometer deflects full scale when it carries a current of 0.500 mA, what resistance must be connected in series with the galvanometer if this combination is to be used as a voltmeter having a full-scale deflection for a potential difference of 1.00 V?

10. (a) Calculate the current through the 6.0-V battery in Figure P28.10. (b) Determine the potential difference between points *a* and *b*.

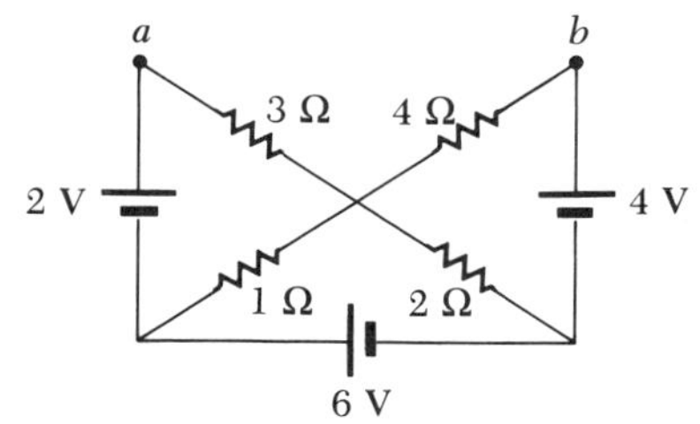

Fig. P28.10

11. In Figure P28.76, calculate (a) the current in the 4.0-Ω resistor, (b) the potential difference between points *a* and *b*, (c) the terminal potential difference of the 4.0-V battery, and (d) the thermal energy expended in the 3.0-Ω resistor during 10 min of operation.

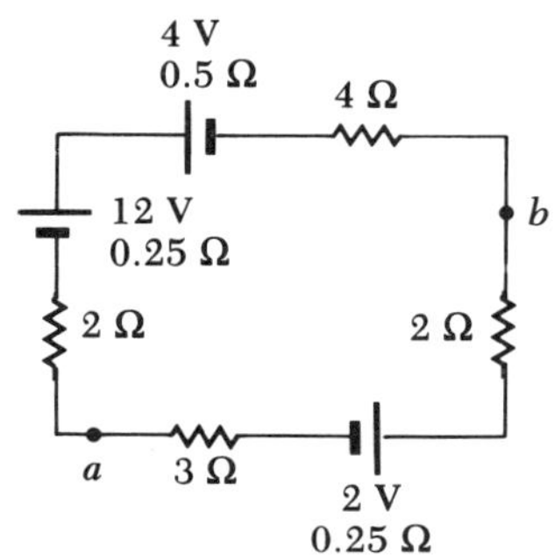

Fig. P28.11

Chapter 29

1. An alpha particle (which is the nucleus of a helium atom) is moving with a northward velocity of 3.8×10^5 m/s in a region where the magnetic field is 1.9 T and points horizontally to the east. What are the magnitude and direction of the magnetic force on this particle?

2. An electron is accelerated through 2400 V and then enters a region where there is a uniform 1.7-T magnetic field. What are (a) the maximum and (b) the minimum values of the magnetic force this charge experiences?

3. The magnetic field over a certain region is $\mathbf{B} = (4.0\mathbf{i} - 11\mathbf{j})$ T. An electron moves in the field with a velocity $\mathbf{v} = (-2.0\mathbf{i} + 3.0\mathbf{j} - 7.0\mathbf{k})$ m/s. Write out in unit-vector notation the force exerted on the electron by the magnetic field.

4. Calculate the magnitude of the force per unit length exerted on a conductor carrying a current of 22 A in a region where a uniform magnetic field has a magnitude of 0.77 T and is directed perpendicular to the conductor.

5. A current $I = 15$ A is directed along the positive x axis in a wire oriented perpendicular to a magnetic field. The wire experiences a magnetic force per unit length of 0.63 N/m in the negative y direction. Calculate the magnitude and direction of the magnetic field.

Chapter 29

6. At a certain location, the Earth has a magnetic field of 6.0×10^{-5} T, pointing 75° below the horizontal in a north-south plane. A 10-m-long straight wire carries a 15-A current. (a) If the current is directed horizontally toward the east, what are the magnitude and direction of the magnetic force on the wire? (b) What are the magnitude and direction of the force if the current is directed vertically upward?

7. When a small bar magnet is suspended in a uniform 0.25-T magnetic field, it experiences a maximum torque of magnitude 4.6×10^{-3} N·m. Calculate its magnetic moment.

8. A circular coil of 100 turns has a radius of 0.025 m and carries a current of 0.10 A while in a uniform magnetic field of 1.5 T. How much work must be done to rotate the coil from a position where the magnetic moment is parallel to the field to a position where the magnetic moment is antiparallel to the field?

9. A 2.0-keV electron moving perpendicular to the Earth's magnetic field of 50 mT has a circular trajectory. Determine (a) the radius of the trajectory and (b) the time required for the electron to complete one circle. (c) Show that your answer to (b) is consistent with the cyclotron frequency of the electron.

10. What is the magnitude of the magnetic field required to constrain a 725-eV electron to a circular path of radius 0.500 m?

11. A beam of protons (all with velocity v) emerge from a particle accelerator and are deflected in a circular arc of radius 0.45 m by a uniform magnetic field of 0.80 T directed perpendicular to the motion of the protons. (a) Determine the speed of the protons. (b) What time is required for a particular proton to deflect through 90°? (c) What is the energy of the protons?

12. (a) What must be the magnetic field strength in a 60.0-in.-diameter cyclotron if it is to accelerate protons to 10.5 MeV? (b) At what frequency must the cyclotron operate? (c) If the frequency is maintained at the value found in (b), to what value must the magnetic field strength be altered if the cyclotron is to accelerate deuterons? (d) After the cyclotron is adjusted to accelerate deuterons, what is the maximum kinetic energy of the deuterons produced?

13. Suppose the conductor shown in Figure P29.13 is copper and is carrying a current of 10 A in a magnetic field of 0.50 T. The width of the conductor d is 1.0 cm and the thickness t is 1.0 mm. Find the Hall potential across its width.

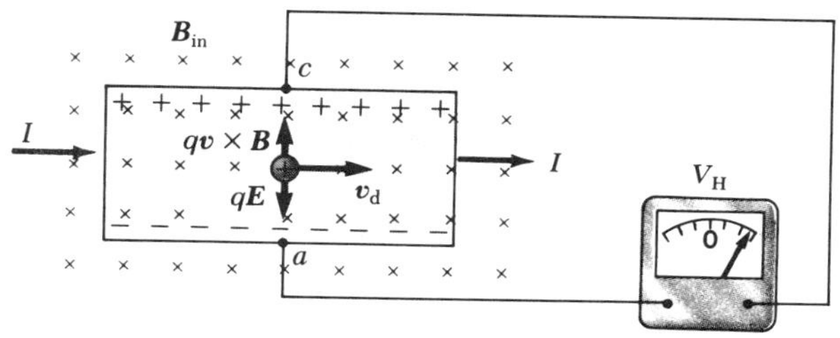

Figure P29.13

14. A proton having velocity $\mathbf{v} = 2.00 \times 10^8\mathbf{i}$ m/s enters a uniform 1.00-T magnetic field and then leaves the field with a velocity $\mathbf{v} = -2.00 \times 10^8\mathbf{j}$ m/s. Determine (a) the direction of the field, (b) the radius of curvature of the proton's path while in the field, (c) the distance traveled by the proton while in the field, and (d) the time spent by the proton in the field.

15. A thin rod of length L is made of a nonconducting material and carries a uniform charge per unit length λ. The rod is rotated with angular speed ω about an axis through its center, perpendicular to the length of the rod. Show that the magnetic dipole moment is $\omega\lambda L^3/24$. (*Hint:* consider the charge dq located in the element dx a distance x from the axis.)

Chapter 29

16. A Bainbridge mass spectrometer is used to examine the isotopes of uranium. Ions in the beam emerge from the velocity selector with a speed equal to 3×10^5 m/s and enter a uniform magnetic field of 0.6 T directed perpendicular to the velocity of the ions. What is the distance between the impact points formed on the photographic plate by singly charged ions of ^{235}U and ^{238}U?

17. A cyclotron designed to accelerate deuterons has a magnetic field that has a uniform magnitude of 1.5 T over a region of radius 0.45 m. If the alternating potential applied between the dees has a maximum value of 15 kV, what time is required for the deuterons to acquire maximum energy?

Chapter 30

1. How many turns should be in a flat circular coil of radius 0.1 m in order for a current of 10 A to produce a magnetic field of 3×10^{-3} T at its center?

2. A circular coil of wire of radius 10.0 cm has 100 turns. The current in the coil is 0.300 A. (a) Calculate the magnetic field at the center of the coil. (b) How far above or below the center are the points where the magnetic field is half its value at the center?

3. Recalling that the current density $J = nqv_d$ (Eq. 27.6), show that the Biot-Savart law can be written

$$d\mathbf{B} = \frac{\mu_0}{4\pi}\frac{q\mathbf{v}_d \times \hat{\mathbf{r}}}{r^2} n \, dV$$

where dV is the volume element of the conductor and the drift velocity $\mathbf{v}_d$ is defined in Chapter 27.

4. A toroid (Fig. P30.4) has 400 turns on a torus that has inner radius a = 4.0 cm and outer radius b = 6.0 cm. Calculate the magnitude of the magnetic field at a point midway between the inner and outer walls of the torus when there is a current of 0.50 A in the windings.

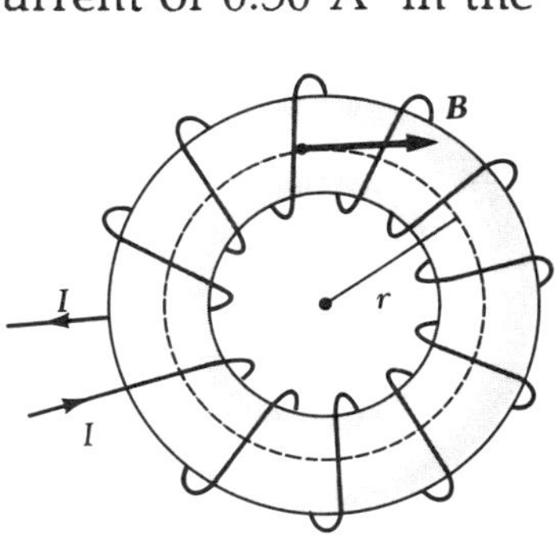

Figure P30.4

5. A cylindrical conductor of radius R = 2.5 cm carries a current I = 2.5 A along its length; this current is uniformly distributed throughout the cross section of the conductor. Calculate the magnetic field midway along a radius (that is, at $r = R/2$).

Chapter 30

6. For the conductor described in Problem 28, find the distance beyond the surface of the conductor at which the magnitude of the magnetic field has the same value as at $r = R/2$.

7. A short solenoid with a length of 10.0 cm and having a radius of 5.0 cm consists of 200 turns of fine wire that carries a current of 15 A. What is the magnetic field strength at the center of the solenoid? For the same number of turns per unit length, what value of B would result for $\ell \longrightarrow \infty$?

8. A circular loop of wire of radius R is placed in a uniform magnetic field **B** and is then spun at constant angular speed ω about an axis through its diameter. Determine the magnetic flux through the loop as a function of time if the axis of rotation is (a) perpendicular to **B** and (b) parallel to **B**.

9. Show that the product of magnetic field strength H and magnetic flux density B has SI units of J/m^3.

10. The core of a solenoid having 250 turns per meter is filled with a material of unknown composition. When the current in the solenoid is 2 A, measurements reveal that the magnetic field within the core is 0.13 T. Determine the magnetic susceptibility of the material and classify it magnetically.

11. The density of a specimen of a suspected new element is determined to be 4.15 g/cm^3. The saturation magnetism of the material is found to be 7.6×10^4 A/m, and the measured atomic magnetic moment is 1.2 Bohr magnetons. Calculate the expected value of the atomic weight of the element based on these values.

Chapter 31

1. A square, single-turn coil 0.20 m on a side is placed with its plane perpendicular to a constant magnetic field. An emf of 18 mV is induced in the coil when its area decreases at a rate of 0.10 m^2/s. What is the magnitude of the magnetic field?

2. The plane of a rectangular coil of dimensions 5.0 cm by 8.0 cm is perpendicular to a magnetic field **B**. If the coil has 75 turns and a total resistance of 8.0 W, at what rate must the magnitude of **B** change in order to induce a current of 0.10 A in the coil?

3. A flat loop of wire of 10 turns, each of area 14 cm^2, is perpendicular to a magnetic field whose magnitude changes in time according to $B = (0.50\ \text{T})\sin(60\pi t)$. What is the induced emf in the loop as a function of time?

4. A circular loop with a radius R consists of N tight turns of wire and is penetrated by an external magnetic field directed perpendicular to the plane of the loop. The magnitude of the field in the plane of the loop is $B = B_0(1 - r/2R)\cos\omega t$, where r is measured from the center of the loop, as in Figure P31.4. Determine the induced emf in the loop.

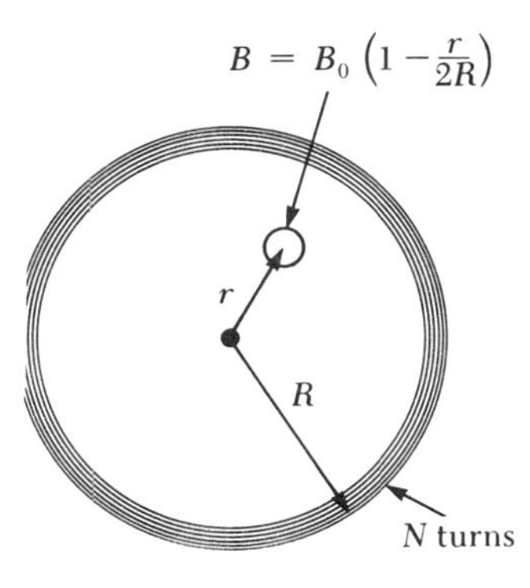

Fig. P31.4

Chapter 31

5. An average emf of 30 mV is induced in a small circular coil of 600 turns and 4.0-cm diameter under the following condition: the coil is rotated in a uniform magnetic field in 0.050 s from a position where the plane of the coil is parallel to the field to a position where the plane of the coil is at an angle of 45° to the field. What is the value of B in the region where the measurement is made?

6. If a constant force of 1.00 N moves the bar in Figure P31.6 at 2.00 m/s through a magnetic field **B** that is into the paper, what are (a) the current through an 8.00-Ω resistor R, (b) the rate of energy dissipation in the resistor, and (c) the mechanical power delivered by the force **F**?

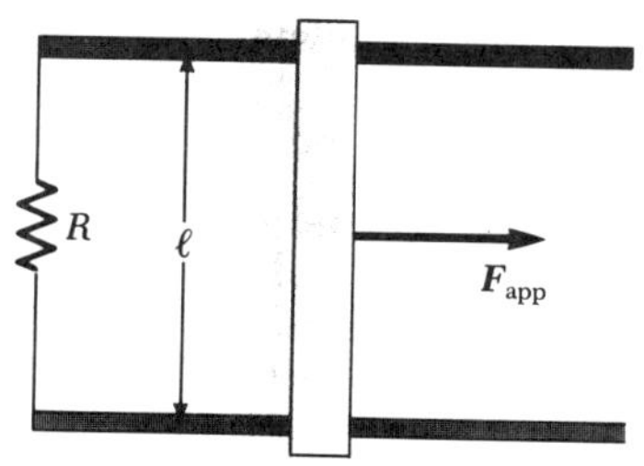

Figure P31.6

7. A small airplane with a wing span of 14.0 m is flying due north at 70.0 m/s over a region where the vertical component of the Earth's magnetic field is 1.20 mT. (a) What potential difference is developed between the wing tips? (b) How would the answer to part (a) change if the plane were flying due east?

8. A rigid, thin conducting rod of length L is mechanically rotated at constant angular speed w about an axis perpendicular to the rod and through its center. If a uniform magnetic field exists parallel to the rotation axis, (a) show that the induced emf between the center of the rod and one of its ends is proportional to L^2. (b) Evaluate the magnitude of this emf for $L = 0.2$ m, $\omega = 60$ rad/s, and $|\mathbf{B}| = 1.2$ T.

Chapter 31

9. A 200-turn circular coil of radius 10 cm is located in a uniform magnetic field of 0.80 T such that the plane of the coil is perpendicular to the direction of the field. The coil is rotated at a constant rate through 90° in 1.5 s, so that its plane is finally parallel to the direction of the field. (a) Calculate the average emf induced in the coil. (b) What is the instantaneous emf at the moment the plane of the coil makes an angle of 45° with the magnetic field?

10. A solenoid has a radius of 2.0 cm and 1000 turns/meter. The current varies with time as $I = 3.0e^{0.2t}$, where I is in amperes and t is in seconds. Calculate the electric field 5.0 cm from the axis of the solenoid at $t = 10$ s.

11. A 400-turn circular coil of radius 15 cm is rotating about an axis perpendicular to a magnetic field of 0.020 T. What angular speed produces a maximum induced emf of 4.0 V?

12. The coil of a simple ac generator develops a sinusoidal emf of maximum value 90.4 V and frequency 60.0 Hz. If the coil is 10.0 cm by 20.0 cm and rotates in a 1.00-T magnetic field, how many turns are in the winding?

13. A solenoid wound with 2000 turns/meter is supplied with current that varies in time according to $I = 4.0 \sin(120\pi t)$, where I is in amperes and t is in seconds. A small coaxial circular coil of 40 turns and radius $r = 5.0$ cm is located inside the solenoid near its center. (a) Derive an expression that describes the manner in which the emf in the small coil varies in time. (b) At what average rate is energy dissipated in the small coil if the windings have a total resistance of 8.0 Ω?

Chapter 31

14. Magnetic field values are often determined by using a device known as a search coil. This technique depends on measuring the total charge through a coil while the magnetic flux through the coil changes either because of the motion of the coil or because of a change in the value of *B*. (a) Show that if the flux through the coil changes from Φ_1 to Φ_2, the charge through the coil between t_1 and t_2 is given by $Q = N(\Phi_2 - \Phi_1)/R$, where *R* is the resistance of the coil and associated circuitry (galvanometer) and *N* is the number of terms. (b) Calculate *B* when a 100-turn coil of resistance 200 Ω and cross-sectional area 40.0 cm^2 produces the following results: A total charge of 5.00×10^{-4} C passes through the coil when it is rotated in a uniform field from a position where the plane of the coil is perpendicular to the field into a position where the coil's plane is parallel to the field.

Chapter 32

1. What is the inductance of a 510-turn solenoid that has a radius of 8.0 cm and a length of 1.4 m?

2. Show that the following expressions have the same units:

$$L = \frac{N\Phi_m}{I} \quad \text{and} \quad L = -\frac{\varepsilon}{dI/dt}$$

3. Three solenoidal windings of 300, 200, and 100 turns are wrapped at well-spaced positions along a cardboard tube of radius 1.0 cm. Each winding extends for 5.0 cm along the cylindrical surface. What is the equivalent inductance of the 600 turns when the three sets of windings are connected in series?

4. Verify by direct substitution that the expression for current given in Equation 32.7 is a solution of Kirchhoff's loop equation for the *RL* circuit as given by Equation 32.6.

5. Calculate the inductance in an *RL* circuit in which $R = 0.500\ \Omega$ and the current increases to one fourth its final value in 1.50 s.

6. A 12-V battery is connected in series with a resistor and an inductor. The circuit has a time constant of 500 μs, and the maximum current is 200 mA. What is the value of the inductance?

7. Show that the inductive time constant t has SI units of seconds.

8. Consider the circuit shown in Figure P32.8, taking $\mathcal{E}$ = 6 V, L = 8 mH, and R = 4 Ω. (a) What is the inductive time constant of the circuit? (b) Calculate the current 250 μs after S_1 is closed.
(c) What is the value of the final steady-state current?
(d) How long does it take the current to reach 80% of its maximum value?

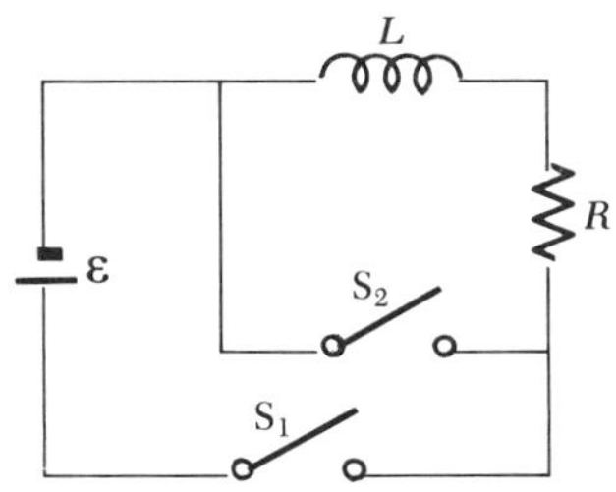

Fig. P32.8

9. Assume that S_1 in Figure P32.8 has been closed long enough for the current to reach its maximum value. If S_1 is now opened and switch S_2 is closed at t = 0, after what time interval is the current in R 25% of the maximum value? In your calculations, take $\mathcal{E}$ = 6 V, L = 24 mH, and R = 10 Ω.

10. In the circuit shown in Figure P32.10 let L = 7.00 H, R = 9.00 Ω, and $\mathcal{E}$ = 120 V. What is the self-induced emf 0.200 s after the switch is closed?

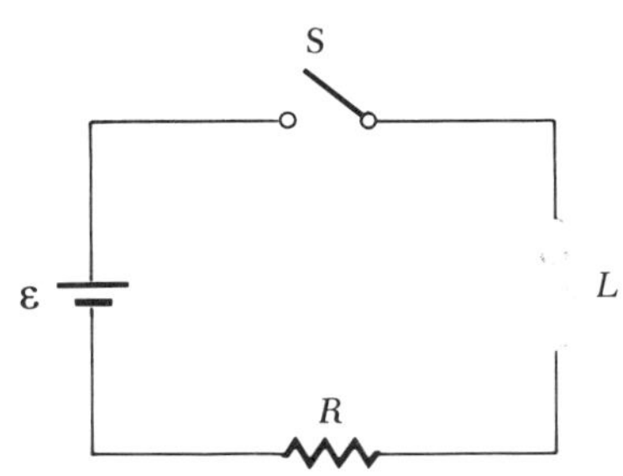

Figure P32.10

Chapter 32

11. The switch in the circuit of Figure P32.11 is thrown closed at $t = 0$. (a) Calculate the rate at which energy is being stored in the inductor after an elapsed time equal to the time constant of the circuit. (b) At what rate is energy being dissipated in the resistor at this time? (c) What is the total energy stored in the inductor at this time?

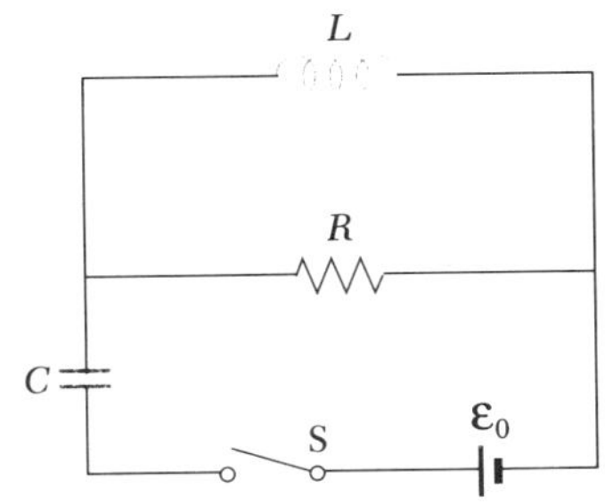

Figure P32.11

12. An RL circuit in which $L = 4.00$ H and $R = 5.00\ \Omega$ is connected to a 22.0-V battery at $t = 0$. (a) What energy is stored in the inductor when the current is 0.500 A? (b) At what rate is energy being stored in the inductor when $I = 1.00$ A? (c) What power is being delivered to the circuit by the battery when $I = 0.500$ A?

13. Two coils A and B located near each other have a mutual inductance $M = 28$ mH. What is the emf induced in coil A as a function of time when the current in coil B is $I = 3.0t^2 - 4.0t + 5.0$, where I is in amperes and t is in seconds?

14. A long solenoid consists of N_1 turns each of radius R_1. A second solenoid, with N_2 turns each of radius R_2, has the same length as the first and lies entirely within the first, with the axes parallel. (a) Assume solenoid 1 carries a current I and compute their mutual inductance. (b) Now assume that solenoid 2 carries the same current I (and solenoid 1 carries no current) and compute their mutual inductance. Do you obtain the same result?

Chapter 32

15. An LC circuit of the type shown in Figure P32.15 has an inductance of 0.57 mH and a capacitance of 15 pF. The capacitor is charged to its maximum value by a 32-V battery. The battery is then removed from the circuit and the capacitor discharged through the inductor. (a) If all resistance in the circuit is neglected, determine the maximum value of the current in the oscillating circuit. (b) At what frequency does the circuit oscillate? (c) What is the maximum energy stored in the magnetic field of the inductor?

Figure P32.15

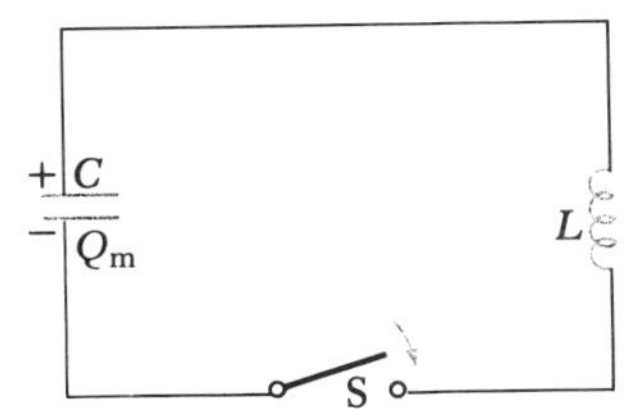

16. An LC circuit carries a current that is oscillating with a period T. If the charge on the capacitor is maximum at t = 0, when does the energy stored in the electric field of the capacitor equal the energy stored in the magnetic field of the inductor? (Express your answer as a fraction of T.)

17. (a) Show that the ratio of the oscillation frequency of a damped LC oscillator (ω_d) to that of an undamped oscillator (ω_0) may be expressed as

$$\frac{\omega_d}{\omega_0} = \sqrt{1 - \frac{R^2C}{4L}}$$

(b) What happens to this ratio if $L < 4R^2C$?

Chapter 32

18. A time-varying current I is applied to an inductance of 5.0 H, as shown in Figure P32.18. Graph the potential at point a relative to that at point b over the interval $t = 0$ to $t = 4.0$ ms. The current arrow indicates the direction of the current.

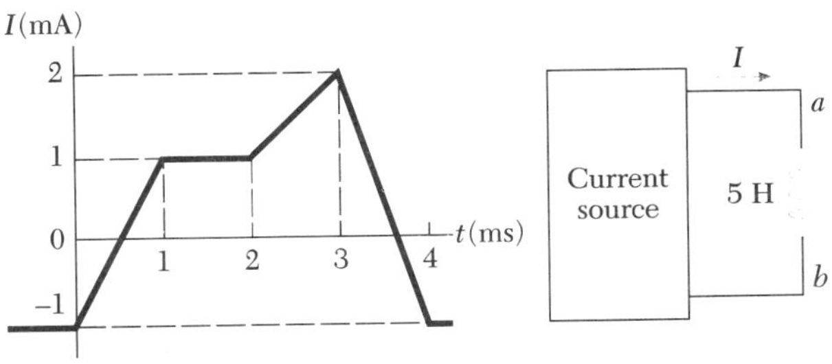

Fig. P32.18

19. A series LR circuit has $L = 0.100$ H and $R = 6.00\ \Omega$. At a certain moment, the current is 5.00 A. (a) What is the potential difference across the inductor? (b) Calculate the time rate of change of current.

20. A flat coil of wire has an inductance of 2.00 H and a resistance of 40.0 Ω. At $t = 0$, a 60.0-V battery is connected to the coil. Find, one second later, (a) the power delivered by the battery, (b) the Joule power developed in the resistance of the windings, and (c) the instantaneous rate at which energy is being stored in the magnetic field.

Chapter 33

1. In the simple ac circuit shown in Figure P33.1, let $R = 60\ \Omega, V_{max} = 100$ V, and $f = 50$ Hz. Assume that the voltage across the resistor $V_R = 0$ when $t = 0$. Calculate (a) the maximum current in the resistor and (b) the angular frequency of the generator.

Figure P33.1

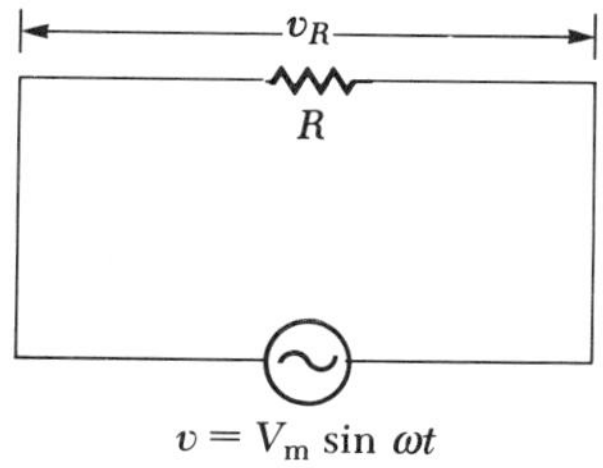

2. Use the values given in Problem 4 for the circuit of Figure 33.1 to calculate the current in the resistor at (a) $t = \frac{1}{75}$ s and (b) $t = \frac{1}{150}$ s.

3. Show that inductive reactance has the SI unit of ohm.

4. What is the inductance of a coil that has an inductive reactance of 63.0 Ω at an angular frequency of 820 rad/s?

5. Show that the SI unit of capacitive reactance is the ohm.

6. Calculate the capacitive reactance of a 10-μF capacitor connected to an ac generator having an angular frequency of 95π rad/s.

7. A coil with an inductance of 18.1 mH and a resistance of 7 Ω is connected to a variable-frequency ac generator. At what frequency will the voltage across the coil lead the current by 45°?

8. A 400-Ω resistor, an inductor, and a capacitor are in series with a generator. When the frequency is adjusted to $600/\pi$ Hz, the inductive reactance is 700 Ω. What minimum values of capacitance result in a circuit impedance of 910 Ω?

9. Calculate the average power delivered to the series *RLC* circuit described in Problem 25.

Chapter 33

10. Consider the circuit described in Problem 28. (a) What is the power factor of the circuit? (b) What is the rms current in the circuit? (c) What average power is delivered by the source?

11. (a) Compute the quality factor for the circuits described in Problems 26 and 27. (b) Which circuit has the sharper resonance?

12. Consider a series combination of a 10.0-mH inductor, a 100-μF capacitor, and a 10.0-Ω resistor. A 50.0-V (rms) sinusoidal voltage is applied to the combination. Calculate the rms current at (a) the resonant frequency, (b) half the resonant frequency, and (c) double the resonant frequency.

13. Assign the values of *R* and *C* given in Problem 50 to the circuit shown in Figure 33.16 and calculate V_{out}/V_{in} for (a) $\omega = 300\ s^{-1}$ and (b) $\omega = 7 \times 10^5\ s^{-1}$.

14. Consider an ideal transformer with N_1 primary and N_2 secondary windings. Show that a step-up transformer (one with $N_2 > N_1$) reduces the current in the output by a factor of N_1/N_2.

15. A particular transformer is 95% efficient and has twice as many secondary windings as primary windings. If the primary windings carry a 5.0-A current at an rms voltage of 120 V, what are the secondary current and rms voltage?

16. The primary current of an ideal transformer is 8.5 A when the primary voltage is 77 V. Calculate the voltage across the secondary when a current of 1.4 A is delivered to a load resistor.

17. A 2.5-V (rms), 100-Hz generator is connected in series with a 100-nF capacitor and a 2500-Ω resistor. Determine (a) the impedance of the circuit and (b) the rms current in the circuit.

Chapter 33

18. An inductor is in series with an 80-W resistor, and the combination is placed across a 110-V (rms), 60-Hz power source. If the resistor dissipates 50 W of power, find the inductance of the inductor.

Chapter 34

1. Verify that Equation 34.12 gives c with dimensions of length per unit time.

2. Calculate the maximum value of the magnetic field in a region where the maximum value of the electric field is 5.3 mV/m.

3. Verify that the following equations for E and B are solutions of Equations 34.6 and 34.7:

$$E = \frac{A}{\sqrt{\varepsilon_0 \mu_0}} e^{a(x-ct)} \qquad B = Ae^{a(x-ct)}$$

4. The amplitude of the magnetic field in an electromagnetic wave is $B_{max} = 4.1 \times 10^{-8}$ T. What is the average intensity of this wave?

5. At a particular point, the electric field in an electromagnetic wave has the instantaneous value $E = 240$ V/m. Compute the instantaneous magnitudes of the following quantities at the same point and time: (a) the magnetic field; (b) the Poynting vector; (c) the energy density of the electric field; (d) the energy density of the magnetic field.

6. A long wire has a radius of 1 mm and a resistance per unit length of 2.00 W /m. Determine the current required if the Poynting vector at the surface of the wire equals 2.68×10^3 W/m^2.

7. A plane electromagnetic wave of intensity 6.0 W/m^2 strikes a small pocket mirror of area 40 cm^2 held perpendicular to the approaching wave. (a) What momentum does the wave transfer to the mirror each second? (b) Find the force that the wave exerts on the mirror.

Chapter 34

8. (a) Assuming that the Earth absorbs all the sunlight incident upon it, find the total force that the Sun exerts on the Earth due to radiation pressure. (b) Compare this value with the Sun's gravitational attraction.

9. Two hand-held radio transceivers with dipole antennas are separated by a large fixed distance. Assuming a vertical transmitting antenna, what fraction of the maximum received power appears in the receiving antenna when it is inclined from the vertical by (a) 15°; (b) 45°; (c) 90°?

10. The eye is most sensitive to light having a wavelength of 5.5×10^{-7} m, which is in the green-yellow region of the electromagnetic spectrum. What is the frequency of this light?

11. Compute the frequency of the following electromagnetic waves: (a) microwaves ($\lambda = 1$ cm); (b) infrared radiation ($\lambda = 1$ mm); (c) yellow light ($\lambda = 580$ nm); (d) ultraviolet light ($\lambda = 100$ nm); and (e) x-rays ($\lambda = 1$ pm).

12. Show that the instantaneous value of the Poynting vector has a magnitude

$$S = \frac{c}{2}(\varepsilon_0 E^2 + \mu_0 H^2)$$

13. The intensity of solar radiation at the top of the Earth's atmosphere is 1350 W/m^2. Assuming that 60% of the arriving solar energy reaches the earth's surface and assuming that you absorb 50 percent of the incident energy, estimate the amount of solar energy you absorb in a 60-min sunbath.

Chapter 34

14. Monoenergetic x-rays move through a material with a speed of $0.95c$. The flux on a surface perpendicular to the x-ray beam is 10^{13} photons/$m^2 \cdot s$. (a) Calculate the density of photons (number per unit volume) in the material. (b) If each photon has an energy of 8.88 keV, what is the energy density within the material?

15. Assuming that the antenna of a 10.0-kW radio station radiates electromagnetic energy uniformly in all directions, compute the maximum value of the magnetic field 5.00 km from the antenna, and compare this value with the magnetic field of the Earth at the surface of the Earth.

Chapter 35

1. Experimenters at the National Bureau of Standards have made precise measurements of the speed of light using the property of electromagnetic waves that in vacuum the phase velocity of the waves is $c = \sqrt{1/\mu_0\varepsilon_0}$, where

   μ_0 (permeability constant) = $4\pi \times 10^{-7}$ N·s^2/C^2 and

   ε_0 (permittivity constant) = 8.854×10^{-12} C^2/N·m^2

   What value (to four significant figures) does this give for the speed of light in vacuum?

2. In Galilio's attempt to determine the speed of light, his observers were located on hilltops 9 km apart. What time interval between uncovering his lantern and seeing the light from his observer's lantern would Galileo require in order to obtain a value for the speed of light?

3. Light is incident on the interface between air and polystyrene at an angle of 53.0°. The incident ray, initially traveling in air, is partially refracted and partially reflected at the surface. What is the angle between the refracted and the reflected ray?

4. Light of vacuum wavelength λ_0 has a wavelength of 438 nm in water and 390 nm in benzene. What is the index of refraction of water relative to benzene at λ_0?

5. A cylindrical tank with an open top has a diameter of 3.00 m and is completely filled with water. When the setting Sun reaches an angle of 28.0° above the horizon, sunlight ceases to illuminate the bottom of the tank. How deep is the tank?

6. Calculate the index of refraction of an equiangular prism for which the angle of minimum deviation is d_{min} = 37°.

Chapter 35

7. A crown glass prism has an apex angle of 15.0°. What is the angle of minimum deviation of this prism for light of wavelength 525 nm. crown glass At this wavelength, crown glass has an index of refraction of 1.52

8. For a particular prism and wavelength, $n = 1.62$. Compare the values found for the angle of minimum deviation in this prism when using the approximation of Problem 25 and the exact form given by Equation 35.9 when (a) $\Phi = 30°$ and (b) $\Phi = 10°$.

9. A fish in a pond is located 15 m from shore. Below what depth is the fish unable to see a small rock at the water's edge?

10. A narrow beam of light is incident from air onto a glass surface of index of refraction 1.56. Find the angle of incidence θ for which the corresponding angle of refraction is $\theta/2$. (*Hint:* Use the trigonometric identity $\sin 2\theta = 2 \sin \theta \cos \theta$)

11. A light beam is incident on a water surface from air. What is the maximum possible value for the angle of refraction?

12. A point source of light at point A is in air a distance h_1 above a water surface, and point B is a depth h_2 below the water surface. Show that the time required for light to travel from A to B is

$$t = \frac{h_1 \sec\theta_1}{c} + \frac{h_2\, n \sec\theta_2}{c}$$

where n is the index of refraction of water, θ_1 is the angle of incidence, and θ_2 is the angle of refraction.

Chapter 35

13. One technique for measuring the angle of a prism is shown in Figure P35.13. A beam of light is directed on the angle such that the rays of the beam reflects from opposite sides. Show that the angular separation of the two rays is given by$B = 2A$.

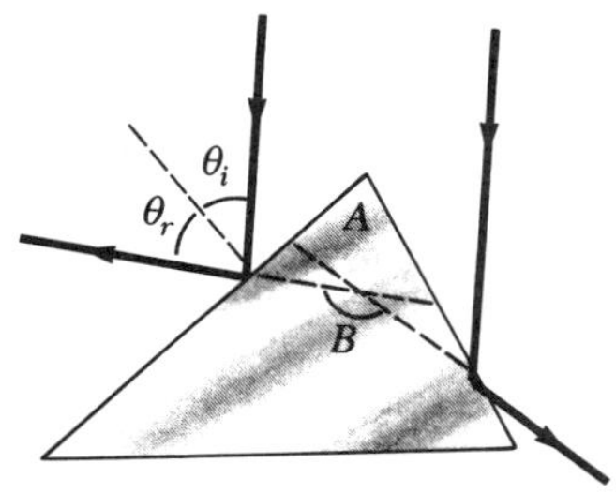

Fig. P35.13

14. Figure P35.14 shows a top view of a square enclosure. The inner surfaces are flat mirrors. A ray of light enters a small hole in the center of one mirror. (a) At what angle θ must the ray enter in order to exit through the hole after being reflected once by each of the other three mirrors? (b) Are there other values of θ for which the ray can exit after multiple reflections? If so, make a sketch of one of the ray's paths.

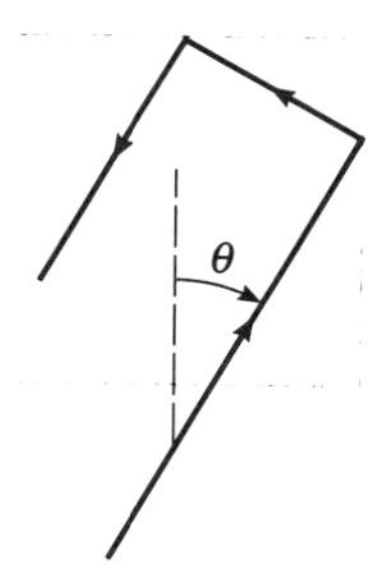

Fig. P35.14

Chapter 36

1. Two flat mirrors, A and B, are in contact along one edge, and the planes of the two mirrors are at an angle of 45° with respect to each other (Fig. P36.1). A point object is placed at P along the bisector of this angle. Make a sketch similar to Figure P36.2 to a suitable scale and locate graphically (a) the image of P in A and in B. (b) Label the images found in (a) P'_A and P'_B, respectively, and locate the image of P'_A in B and the image of P'_B in A.
(c) Determine the total number of images for the arrangement.

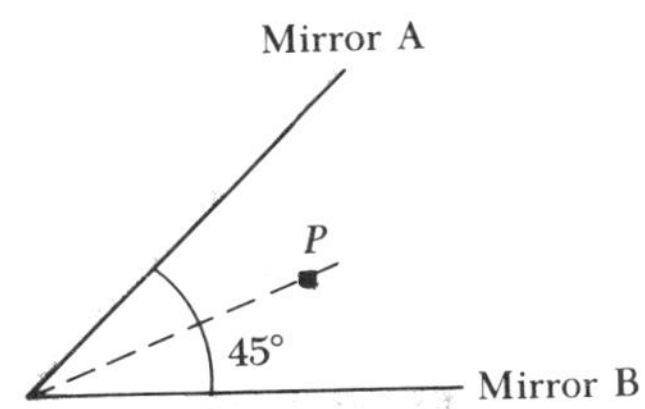

Fig. P36.1

2. A convex mirror has a focal length of – 20.0 cm. Determine an object location for which the image is one half the size of the object.

3. Calculate the image position and magnification for an object placed (a) 20 cm and (b) 60 cm in front of a convex mirror of focal length – 40 cm. (c) Use ray diagrams to locate these images.

4. Use a ray diagram to demonstrate that the image of a real object placed in front of a spherical mirror is always virtual and upright when $p < |f|$.

5. An object is located at the zero end of a meter stick. A concave mirror located at the 100-cm end forms an image at the 70-cm position. A convex mirror at the 60-cm position forms a final image at the 10-cm point. What is the radius of curvature of the convex mirror?

6. Calculate the image positions in Problem 18 if the end of the rod is a concave surface of radius 8.0 cm.

7. A thin-walled, hollow convex lens is immersed in a tank of water. The lens has $R_1 = 20.0$ cm and $R_2 = 30.0$ cm. Calculate the focal length of this "air lens" surrounded by water ($n = 1.33$). Use the derivation of Equation 36.11 as a guide.

8. What is the unaided near point for a person required to wear lenses with a power of + 1.5 diopters to read at 25 cm?

9. Assume that the camera shown in Figure P36.9 has a fixed focal length of 6.5 cm and is adjusted to properly focus the image of a distant object. By how much and in what direction must the lens be moved in order to focus the image of an object at a distance of 2.0 m?

Figure P36.9

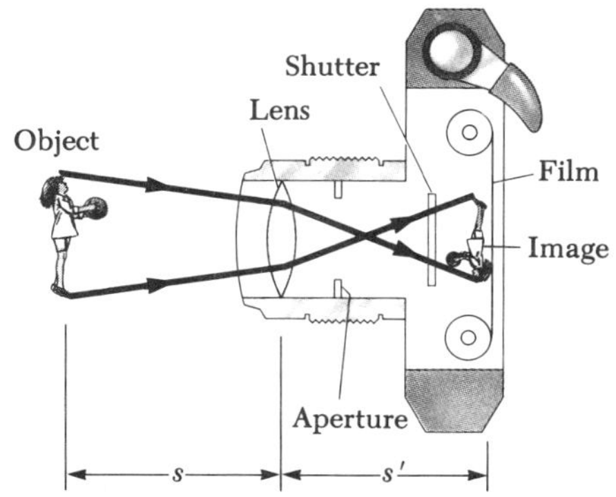

Chapter 36

10. Figure P36.10 illustrates the case of a farsighted person who can focus clearly on objects that are more distant than 90 cm from the eye. Determine the lens power that enables this person to read comfortably at a normal near point of 25 cm.

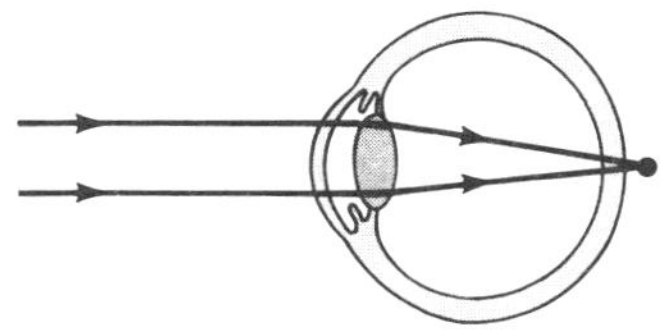

Figure P36.10

11. The eye of a nearsighted person is illustrated in Figure P36.11. In this case, the person cannot focus clearly on objects that are more distant than 200 cm from the eye. Determine the lens power that enables this person to see distant objects clearly.

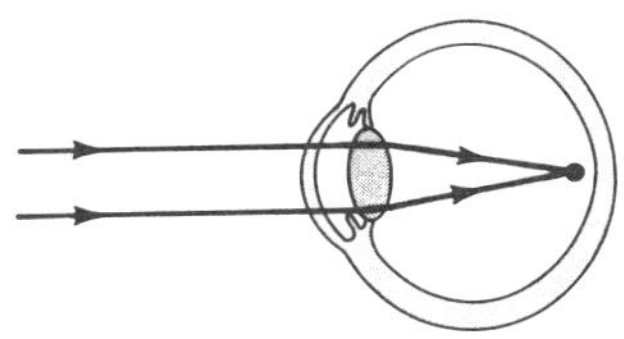

Figure P36.11

12. An object is located in front of a screen. A thin lens placed between object and screen produces a sharp image on the screen when the lens is in either of two positions that are 10 cm apart. The sizes of images in the two situations are in the ratio 3:2. What are (a) the focal length of the lens and (b) the distance from screen to object?

13. Show that, for a converging lens, unit magnification ($m = -1$) results when $p = 2f$. Under what condition is $m = +1$ approached?

14. Construct a graph of angular magnification m as a function of p for (a) a converging lens and (b) a diverging lens.

15. A concave mirror has a radius of 40.0 cm. (a) Calculate the image distance q for an arbitrary object distance p. (b) Obtain values of q for object distances of 5.0 cm, 10.0 cm, 40.0 cm, and 60.0 cm. (c) Make a plot of q versus p using the results of (b).

16. Find the object distances (in terms of f) of a thin converging lens of focal length f if (a) the image is real and the image distance is $4f$ and (b) the image is virtual and the image distance is $3f$. (c) Calculate the magnification of the lens for parts (a) and (b).

Chapter 37

1. Yellow light (λ = 587.5 nm) falls on a plane containing parallel slits 0.20 mm apart. A screen is located so that the second bright band in the interference pattern is at a distance equal to 10 slit spacings from the central maximum. What is the distance between slits and screen?

2. A narrow slit is cut into each of two overlapping opaque squares. The slits are parallel, and the distance between them is adjustable. Monochromatic 600-nm light illuminates the slits, and an interference pattern is formed on a screen 80 cm away. The third dark band is 1.2 cm from the central bright band. What is the distance between the center of the central bright band and the center of the first bright band on either side of the central band?

3. In deriving Equation 37.5, we assumed $\sin\theta \approx \tan\theta$ which is true only when $d >> \lambda$. In the arrangement of Figure P37.3, let L = 40 cm, d = 0.5 mm, and λ = 656.3 nm. Under the assumption $\sin\theta \approx \tan\theta$, the ninth-order dark bands are located 4.4631 mm on either side of the central maximum. What percent error is introduced by the assumption $d >> \lambda$?

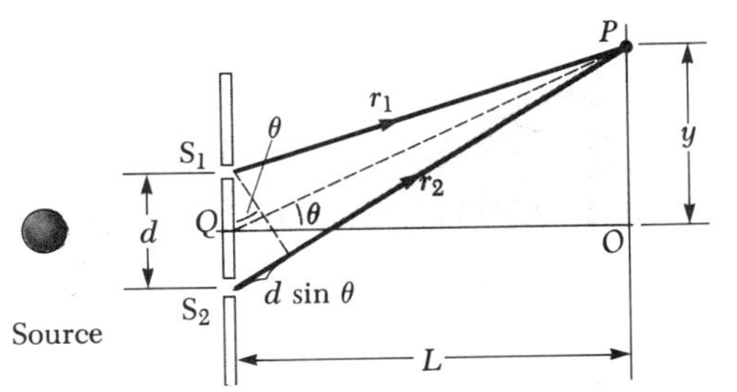

Figure P37.3

4. Light of wavelength 546 nm produces a Young's interference pattern in which the second-order minimum is along a direction that makes an angle of 18 minutes of arc relative to the direction to the central maximum. What is the distance between the parallel slits?

5. In Figure P37.5, let $L = 140$ cm and $y = 8.0$ mm. Find the value of the ratio d/λ for which the average intensity at point P is 60% of the maximum intensity.

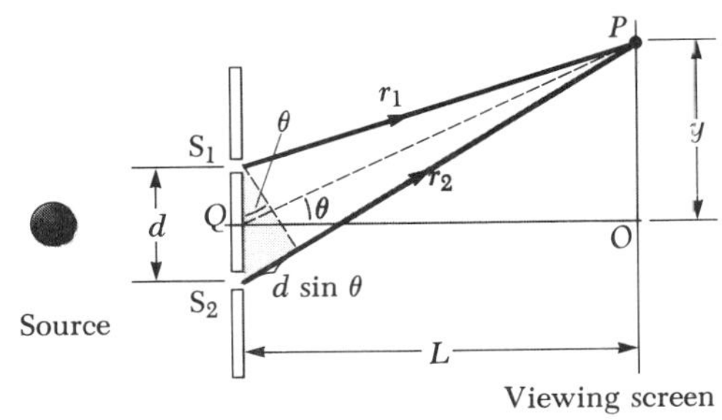

Figure P37.5

6. A soap bubble appears green when exposed to normally incident white light. Determine the minimum thickness of the soap film that can produce the observed effect. Take the wavelength of green light to be 500 nm. The index of refraction of the film is 1.33.

7. A microwave version of the Michelson interferometer uses 4.00-cm waves. The detector receiving the two interfering microwave beams is sensitive to phase differences corresponding to path differences of 0.100l. What is the smallest displacement that can be detected with the interferometer?

8. In Example 37.1, the double-slit arrangement produced fringe separations of 2.2 cm for $\lambda = 560$ nm. Calculate the fringe separation for this same arrangement if the apparatus is submerged in a tank containing a 30% sugar solution ($n = 1.38$).

9. Two sinusoidal phasors of the same amplitude A and frequency ω have a phase difference ϕ. Calculate the resultant amplitude of the two phasors both graphically and analytically if ϕ equals (a) 0, (b) 60°, (c) 90°.

Chapter 37

10. A thin film of oil (n = 1.38) having a thickness d floats on water. (a) Show that the equation

$$\lambda = \frac{2nd}{(m + \frac{1}{2})} \qquad m = 0, 1, 2, \ldots$$

gives the wavelengths that exhibit constructive interference for reflected light. (b) What color is the film at points where d = 300 nm? (The film is viewed from above in white light.)

11. A fringe pattern is established in the field of view of a Michelson interferometer using 580-nm light. A sheet of transparent material 2.5 mm thick is placed in front of one of the mirrors perpendicular to the incident and reflected light beams. An observer counts a fringe shift of six dark fringes. What is the index of refraction of the sheet?

12. A soap film such as that in Problem 66 is viewed in 589-nm light with near-normal incidence. Interference fringes with a uniform spacing of 4.5 mm are observed. What is the variation in thickness of the film per centimeter of vertical distance?

Chapter 38

1. In Figure P38.1, let the slit width a = 0.500 mm and assume incident monochromatic light of wavelength 460 nm. (a) Find the value of θ corresponding to the second-order dark fringe. (b) If the screen is 120 cm in front of the slit, what is the distance y from the center of the central maximum to the second dark fringe?

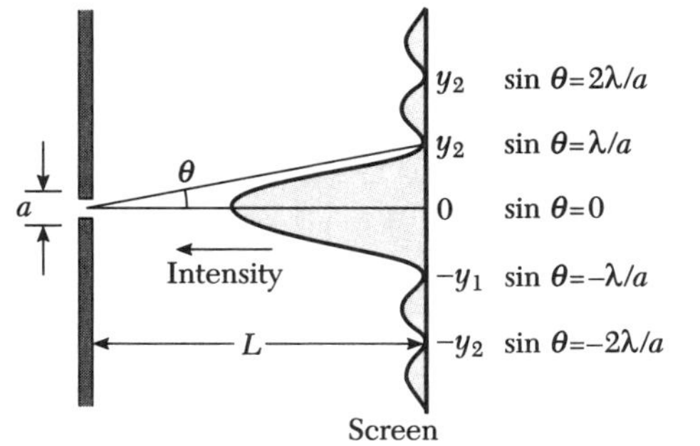

Figure P38.1

2. In the calculation of Example 38.2, we assumed that the first- and second- order maxima in the single-slit Fraunhofer diffraction pattern are at the locations $\beta/2 = 3\beta/2$ and $\beta/2 = 5\beta/2$. (a) Prove that these maxima are really at locations given by $\tan(\beta/2) = \beta/2$. (b) Find the first three positive solutions for $\beta/2$ to five digits.

3. What minimum distance between two points permits them to be resolved at 1.0 km (a) by a terrestrial telescope with a 6.5-cm-diameter objective (assume λ = 550 nm) and (b) by the unaided eye (assume a pupil diameter of 2.5 mm)?

4. The Hubble space telescope, now in Earth orbit, has a mirror 2.4 m in diameter, ground to the wrong shape. If corrected, what is the best angular resolution it could achieve for visible light of wavelength 500 nm?

Chapter 38

5. Collimated light from a hydrogen discharge tube is incident normally on a diffraction grating. The incident light includes four wavelength components: $\lambda_1 = 410.1$ nm, $\lambda_2 = 434.0$ nm, $\lambda_3 = 486.1$ nm, and $\lambda_4 = 656.3$ nm. There are 410 lines/mm in the grating. Calculate the angle between (a) λ_1 and l_4 in the first-order spectrum and (b) λ_1 and λ_3 in the third-order spectrum.

6. Monochromatic light is incident on a grating that is 75 mm wide and ruled with 50 000 lines. The line is imaged at 32.5° in the second-order spectrum. Determine the wavelength of the incident light.

7. A diffraction grating disperses white light so that the red wavelength $\lambda = 650$ nm appears in the second-order pattern at $\theta = 20°$. (a) Find the so-called grating constant—that is, the number of slits per centimeter. (b) Determine whether or not visible light of the third-order pattern appears at $\theta = 20°$.

8. A diffraction grating that has 2500 rulings/cm is used to examine the sodium spectrum. Calculate the angular separation of the sodium yellow doublet lines (588.995 nm and 589.592 nm) in each of the first three orders.

9. The 501.5-nm line in helium is observed at an angle of 30° in the second-order spectrum of a diffraction grating. Calculate the angular deviation associated with the 667.8-nm line in helium in the first-order spectrum for the same grating.

10. Show why the Bragg condition expressed by Equation 38.11 cannot be satisfied in cases where the wavelength is greater than $2d$ (the length of the unit cell).

Chapter 38

11. The dimension labeled a in Figure P38.11 is the edge length of the unit cell of NaCl. This is twice the distance between adjacent ions, which is the parameter d in Equation 38.11. Calculate d for the NaCl crystal from the following data: density $\rho = 2.164$ g/cm^3, molecular mass $M = 58.45$ g/mol, and Avogadro's number $N_A = 6.022 \times 10^{23}$ atoms/mol.

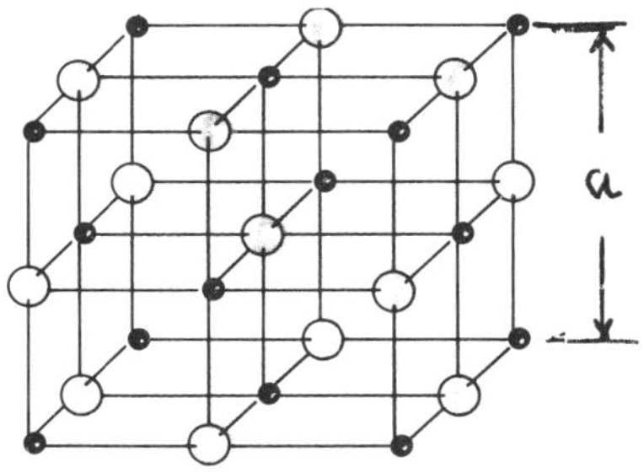

Figure P38.11

12. Unpolarized light passes through two polaroid sheets. The axis of the first is vertical, the second is at 30° to the vertical. What fraction of the initial light is transmitted?

13. Vertically polarized light is passed through three successive polaroid filters. The transmission axes are at 30°, 60°, and 90° to the vertical. What percentage of the light gets through?

14. Light consisting of two wavelength components is incident on a grating. The shorter component has a wavelength of 440 nm. The third-order image of this component is coincident with the second-order image of the longer component. Determine the value of the longer wavelength.

15. Sunlight is incident on a diffraction grating that has 2750 lines/cm. The second-order spectrum over the visible range (400 - 700 nm) is to be limited to 1.75 cm along a screen a distance L from the grating. What is the required value of L ?

Chapter 38

16. Consider a light beam containing two wavelength components whose difference $\Delta\lambda$ is small relative to the mean wavelength λ. When such a beam is incident on a diffraction grating, a useful measure of the angular separation of the maxima corresponding to the two wavelengths is the *dispersion* D, given by $D = d\theta/d\lambda$. (The dispersion of a grating should not be confused with its resolving power R, given by Equations 38.9 and 38.10.)
(a) Starting with Equation 38.8, show that the dispersion can be written

$$D = \frac{\tan\theta}{\lambda}$$

(b) Calculate the dispersion in the third order for the grating described in Problem 65. Give the answer in units of degrees per nanometer.

17. Derive Equation 38.10. Remember that Rayleigh's criterion states that two wavelengths are resolved when the central maximum for one falls on the first minimum for the other.

Chapter 39

1. How fast must a meter stick be moving if its length is observed to shrink to 0.50 m?

2. Each day a clock on a moving spacecraft runs 1 min slower than an identical clock on Earth. What is the speed of the spacecraft? (*Hint:* For $v/c \ll 1$, note that $\gamma \approx 1 + v^2/2c^2$.)

3. An atomic clock moves at 1000 km/h for 1 h as measured by an identical clock on Earth. How many nanoseconds slow is the moving clock after one-hour?

4. A muon formed high in the Earth's atmosphere travels at $v = 0.99c$ for 4.6 km before it decays into an electron, a neutrino, and an antineutrino ($\mu^- \rightarrow e^- + \nu + \bar{\nu}$). How long does the muon live as measured in its reference frame?

5. For what value of v does $\gamma = 1.01$?

6. A golf ball travels at 90 m/s. By what fraction does its relativistic momentum p differ from the classical value mv? That is, find the ratio $(p - mv)/mv$.

7. Show that the kinetic energy K, momentum p, and mass m of a relativistic particle are related by $p^2c^2 = 2Kmc^2 + K^2$.

8. Find the speed at which the relativistic kinetic energy of a particle equals twice the nonrelativistic value.

9. A radium isotope decays by emitting an a particle to a radon isotope according to the scheme ${}^{222}_{88}\text{Ra} \rightarrow {}^{222}_{86}\text{Rn} + {}^{4}_{2}\text{He}$. The masses of the atoms are 226.0254 (Ra), 222.0175 (Rn), and 4.0026 (He). How much energy is released as the result of this decay?

Chapter 39

10. An astronaut is traveling in a space vehicle that has a speed of $0.50c$ relative to the Earth. The astronaut measures her pulse rate to be 75 beats per minute. Signals generated by the astronaut's pulse are radioed to Earth when the vehicle is moving perpendicular to a line connecting the vehicle with an Earth observer. What pulse rate does the Earth observer measure? What would pulse rate would the Earth observer measure if the speed of the space vehicle were increased to $0.99c$?

11. An electron has a speed of $0.75c$. Find the speed of a proton that has (a) the same kinetic energy as the electron and (b) the same momentum as the electron.

12. Determine the speed of recession of the quasar 3C9 if its redshift $\Delta\lambda/\lambda = 2$. Use the relativistic formula for Doppler shift:

$$\frac{\Delta\lambda}{\lambda} + 1 = \sqrt{\frac{c+u}{c-u}}$$

where u = recessional speed and $\Delta\lambda$ is the wavelength shift.

13. The average lifetime of a pi meson in its own frame of reference is 2.6×10^{-8} s. If the meson moves with at $0.95c$, what is (a) its mean lifetime as measured by an observer on Earth and (b) the average distance it travels before decaying, as measured by an observer on Earth?

14. If you travel on a jet plane from New York to Los Angeles (4000 km air distance), at an average speed of 1000 km/h, how much younger are you on arrival than you would have been had you remained in New York ? (*Hint:* Note that T, the time that would have been spent in New York, is extremely close to T_0, the time spent on the plane.)

Chapter 39

15. An antiproton (a particle whose mass is the same as the proton but has opposite charge) can be created in a large synchrotron by an energetic moving proton striking a stationary proton:

$$p^+ + p^+ \rightarrow p^+ + p^+ + p^+ + p^-$$

What minimum kinetic energy must the incoming proton have to produce an antiproton? The rest mass of the p^+ and p^- is 938 MeV/c^2. (*Hint:* The incoming proton has speed v, the outgoing ensemble of four particles leaves with speed u.)

16. Imagine that the Sun collapsed to a sphere of radius R_g such that the work required to remove a small mass m from its surface is equal to its rest energy mc^2. (This radius is called the *gravitational radius* R_g.) Find R_g. (It is believed that the ultimate fate of many stars is to collapse to their gravitational radii or smaller.)

17. *Doppler effect for light.* If a light source moves with a speed v relative to an observer, there is a shift in the observed frequency analogous to the Doppler effect for sound waves. Show that the observed frequency f_0 is related to the true frequency f through the expression

$$f_0 = \sqrt{\frac{c \pm v_s}{c \mp v_s}}\, f$$

where the upper signs correspond to the source approaching the observer and the lower signs correspond to the source receding from the observer. [*Hint:* In the moving frame S′, the period is the proper time interval and is given by $T = 1/f$. Furthermore, the wavelength measured by the observer is $\lambda_0 = (c - v_s)T_0$, where T_0 is the period measured in s.]

Chater 40

1. Consider the mass-spring system described in Example 40.2. If the quantum number n changes by unity, calculate the fractional change in energy of the oscillator.

2. Show that at long wavelengths, Planck's radiation law (Eq. 40.3) reduces to the Rayleigh-Jeans law (Eq. 40.2).

3. When cesium metal is illuminated with 500-nm light, the photoelectrons emitted have a maximum kinetic energy of 0.57 eV. Find (a) the work function of cesium and (b) the stopping potential if the incident light has a wavelength of 600 nm.

4. Lithium has a work function $\phi = 2.3$ eV. (a) Calculate the cutoff frequency for lithium. (b) Calculate the stopping potential for 500-nm light falling on a clean lithium surface.

4. A certain metal has a work function of 3.1 eV. Calculate the stopping potential if the incident light has a wavelength of 270 nm.

5. Ultraviolet light is incident normally on the surface of a certain substance. The work function of the electrons in this substance is 3.44 eV. The incident light has an intensity of 0.055 W/m^2. The electrons are photoelectrically emitted with a maximum speed of 4.2×10^5 m/s. How fast can electrons be emitted from a square centimeter of the surface? Pretend that none of the photons are reflected or heat the surface.

6. Calculate the energy and momentum of a 700-nm photon.

7. What scattering angle shifts the wavelength of a 1.00-nm x-ray by 0.020%?

Chater 40

8. A gamma-ray photon with an energy equal to the rest energy of an electron (511 keV) collides with an electron that is initially at rest. Calculate the kinetic energy acquired by the electron if the photon is scattered 30° from its original line of approach.

9. A 0.03-nm x-ray photon is scattered from a free electron. (a) If the shift in the x-ray wavelength equals the Compton wavelength of the electron, what is the electron's kinetic energy after the collision? (b) What is its speed?

10. A metal target is placed in a beam of 662-keV gamma rays emitted by a radioactive isotope of cesium (^{137}Cs). Find the energy of those photons that are scattered through an angle of 90°. The electrons in the target may be considered as free electrons.

11. X-rays of wavelength 0.12 nm undergo Compton scattering. (a) Find the wavelength of photons scattered at 30°, 60°, 90°, 120°, 150°, and 180°. (b) Find the energy of the scattered electrons corresponding to these scattered x-rays. (c) Which one of the given scattering angles provides the electron with the greatest energy?

12. Use Equation 40.18 to calculate the wavelengths of the first three lines in the Paschen series for hydrogen.

13. Use Equation 40.17 to calculate the wavelengths of the first three lines in the Lyman series for hydrogen.

14. Construct an energy level diagram for the Li^{2+} ion, for which $Z = 3$.

15. What is the energy of the photon that could cause an electronic transition from (a) the $n = 3$ state to the $n = 5$ state and (b) the $n = 5$ state to the $n = 7$ state?

Chater 40

16. (a) Using Equation 40.19 calculate the longest and shortest wavelengths for the Brackett series. (b) Determine the photon energies corresponding to these wavelengths.

17. A photon of initial energy 0.1 MeV undergoes Compton scattering at an angle of 60°. Find (a) the energy of the scattered photon, (b) the recoil energy of the electron, and (c) the recoil angle of the electron.

18. Experiments indicate that a dark-adapted human eye can detect a single photon of visible light. Consider a point source that emits 2.0 W of 555-nm light in all directions. How far away would this source have to be for, on the average, one photon per second to enter an eye having a pupil diameter of 6.0 mm?

Chapter 41

1. Calculate the de Broglie wavelength for a proton moving with a speed of 10^6 m/s.

2. Calculate the de Broglie wavelength of a 75-kg person jogging at 5.0 m/s.

3. Find the de Broglie wavelength of a 0.15-kg ball moving at 20 m/s.

4. Show that the de Broglie wavelength of an electron accelerated from rest through a potential difference V is $\lambda = 1.226/\sqrt{V}$ nm, where V is in volts.

5. An electron has a de Broglie wavelength equal to the diameter of the hydrogen atom. What is the kinetic energy of the electron? How does this energy compare with the ground-state energy of the hydrogen atom?

6. Through what potential difference would an electron have to be accelerated to give it a de Broglie wavelength of 1.00×10^{-10} m?

7. A monoenergetic beam of electrons is incident on a single slit of width 0.50 nm. A diffraction pattern is formed on a screen 20 cm from the slit. If the distance between successive minima of the diffraction pattern is 2.1 cm, what is the energy of the incident electrons?

8. A proton has a kinetic energy of 1 MeV. If its momentum is measured with an uncertainty of 5 percent, what is the minimum uncertainty in its position?

Chapter 41

9. (a) Use the uncertainty principle to estimate the uncertainty in momentum for a particle in a box. (b) Estimate the ground-state energy, and compare your result with the actual ground-state energy.

10. A 5.0-g bead slides freely on a wire 20 cm long. Treating this system as a particle in a one-dimensional box, calculate the value of n corresponding to the state of the bead if it is moving at 0.10 nm per year (that is, apparently at rest).

11. Find the wave functions for the three lowest energy states for a particle in a one-dimensional box. Note that the normalization constant for all three functions in equal to $\sqrt{2/L}$.

12. A particle in the space $-a \le x \le a$ may be represented by either

$$\psi_1 = A \cos\left(\frac{\pi x}{2a}\right) \quad \text{or} \quad \psi_2 = B \sin\left(\frac{\pi x}{a}\right)$$

Using the normalization condition on y, find A and B. (See Problem 37 for a helpful hint.)

13. The wave function of a particle is

$$\psi(x) = A \cos(kx) + B \sin(kx)$$

where A, B, and k are constants. Show that y is a solution of the Schrödinger equation (Eq. 41.15) when the particle is free $(U = 0)$, and find the corresponding energy E of the particle.

Chapter 41

14. Show that the oscillator energies in Equation 41.26 correspond to the classical amplitudes

$$A_n = \sqrt{\frac{\hbar}{m\omega}(2n+1)}$$

15. A two-slit electron-diffraction experiment is done with slits of unequal widths. When only slit 1 is open, the number of electrons reaching the screen per second is 25 times the number of electrons reaching the screen per second when only slit 2 is open. When both slits are open, an interference pattern results in which the destructive interference is not complete. Find the ratio of the probability of an electron arriving at an interference maximum to the probability of an electron arriving at an adjacent interference minimum. (*Hint:* Use the superposition principle.)

16. An electron is trapped in an infinitely deep potential well 0.300 nm in width. (a) If the electron is in its ground state, what is the probability of finding it within 0.100 nm of the left-hand wall? (b) Repeat (a) for an electron in the 99th energy state above the ground state. (c) Are your answers consistent with the correspondence principle? (*Hint:* Use Eq. 41.7.)

17. A particle of mass 2.00×10^{-28} kg is confined to a one-dimensional box of width 1.00×10^{-10} m. For $n = 1$,what are the particle's (a) wavelength, (b) momentum, and (c) ground-state energy?

18. An electron is confined to one-dimensional motion between two rigid walls separated by a distance L. (a) What is the probability of finding the electron within the interval $x = 0$ to $x = L/3$ from one wall if the electron is in its $n = 1$ state? (b) Compare this value with the classical probability.

Chapter 42

1. (a) Determine the quantum numbers ℓ and m_ℓ for the He^+ ion in the $n = 3$ state. (b) What is the energy of this state?

2. (a) Determine the quantum numbers ℓ and m_ℓ for the Li^{2+} ion for the $n = 1$ and $n = 2$ states. (b) What are the energies of these states?

3. Show that the wave function $\psi_{1s}(r)$ given by Equation 42.3 is normalized.

4. If an electron has an angular orbital momentum of 4.714×10^{-34} $kg{\cdot}m^2/s$, what is the orbital quantum number for this state?

5. Calculate the possible values of the z component (the component along the direction of an external magnetic field) of angular momentum for an electron in a d subshell.

6. Like the electron, the nucleus of an atom has spin angular momentum and a corresponding magnetic moment. The z component of the spin magnetic moment for a nucleus is characterized by the nuclear magneton $\mu_n = e\hbar/2m_p$, where m_p is the proton mass. (a) Calculate the value of μ_n in joules per tesla and in electron volts per tesla. (b) Determine the ratio μ_n/μ_B, and comment on your result.

7. Which electronic configuration has the lesser energy and the greater number of unpaired spins: [Kr]$4d^9 5s^1$ or [Kr]$4d^{10}$? Identify this element and discuss Hund's rule in this case. (*Note:* The notation [Kr] represents all the filled shells and subshells for Kr.)

8. What is the shortest x-ray wavelength that can be produced with an accelerating voltage of 10 kV?

Chapter 42

9. A tungsten target is struck by electrons that have been accelerated from rest through a 40-kV potential difference. Find the shortest wavelength of the bremsstrahlung radiation emitted.

10. Find the wavelength of the K_α x-ray line emitted when electrons strike an iron target. Note that because the innermost shells are involved, Z_{eff} is approximately $Z - 1$.

11. Zirconium has two unpaired electrons in the d subshell. (a) What are all possible values of ℓ and s for each electron? (b) What are all possible values of n, m_ℓ, and m_s? (c) What is the electron configuration in zirconium?

12. The force on a magnetic moment μ_Z in a nonuniform magnetic field B_Z is

$$F_Z = \mu_Z \frac{dB_Z}{dz}$$

If a beam of silver atoms travels a horizontal distance of 1.0 m through such a field and each atom has a speed of 100 m/s, how strong must the field gradient dB_Z/dz be in order to deflect the beam 1.0 mm?

Chapter 43

1. Use the data in Table 43.2 to calculate the reduced mass of the NO molecule. Then compute a value for μ using Equation 43.8. Compare the two results.

2. The HF molecule has a bond length of 0.092 nm. (a) Calculate the reduced mass of the molecule. (b) Plot the potential energy versus internuclear separation in the vicinity of $r = 0.092$ nm.

3. The distance between the protons in H_2 is $r = 0.75 \times 10^{-10}$ m. Find (a) the energy of the first rotational state ($J = 1$) and (b) the wavelength of radiation emitted in the transition $J = 1$ to $J = 0$.

4. Find the ionic cohesive energy for LiCl, which has the same crystal structure as NaCl. Take $r_0 = 0.257$ nm, and $m = 7$.

5. Calculate the probability that a conduction electron in copper at 300 K has an energy equal to 99% of the Fermi energy.

6. Find the probability that a conduction electron in a metal has an energy equal to the Fermi energy at 300 K.

7. From the optical absorption spectrum of a certain semiconductor, one finds that the longest wavelength absorbed is 1.85 mm. Calculate the energy gap for this semiconductor.

8. The Fermi energy of copper at 300 K is 7.05 eV. (a) What is the average energy of a conduction electron in copper at 300 K? (b) At what temperature does the average energy of a molecule of an ideal gas equal the energy obtained in (a)?

Chapter 44

1. The penetration depth for lead at 0.0 K is 39 nm. Find the penetration depth in lead at (a) 1.0 K, (b) 4.2 K, and (c) 7.0 K.

2. Find the critical magnetic field in mercury at (a) 1.0 K and (b) 4.0 K.

3. *High-Temperature Superconductor.* Estimate the gap energy E_g for the high-temperature superconductor $YBa_2Cu_3O_{7-\delta}$, which has a critical temperature of 92 K, assuming BCS theory holds.

4. Estimate the area of a ring that would fit one of your fingers, and calculate the magnetic flux through the ring due to the Earth's magnetic field (5.8×10^{-5} T). If this flux were quantized, how many fluxons would the ring enclose?

Chapter 45

1. In Example 45.2, the density of nuclear matter was calculated to be approximately 2.3×10^{17} kg/m^3. What is this density in nucleons/fm^3? in g/cm^3?

2. What would be the gravitational force between two golf balls (each of 4.3-cm diameter), 1.0 m apart, if they were made of nuclear matter?

3. Show that the quantity $e^2/4\pi\varepsilon_0 = 1.440$ MeV·fm. (This is a convenient equivalence for computational purposes.)

4. Using the value of the atomic mass of $^{56}_{26}$Fe given in Table 45.6, find its binding energy. Then compute the binding energy per nucleon and compare your result with Figure 45.7.

5. The neutral atom $^{60}_{28}$Ni has an atomic mass of 59.930 789 u. (a) What is its mass? (b) What is the binding energy per nucleon? Is this a tightly bound or loosely bound nucleus?

6. (a) From Figure 45.8, estimate the electrostatic potential energy due to the Coulomb repulsion between two protons in contact. Model the proton as a uniform sphere of positive charge. (b) Compare this value with the rest energy of an electron.

7. Calculate the binding energy per nucleon for (a) $^{20}_{10}$Ne, (b) $^{40}_{20}$Ca, (c) $^{93}_{41}$Nb, and (d) $^{197}_{79}$Au.

8. A radioactive source having a half-life of 1.04 s has an initial activity of 1.0×10^4 decays/s. Determine the number of unstable nuclei remaining after 1.7 s.

Chapter 45

9. A sample of radioactive material is said to be carrier-free when no stable isotopes of the radioactive element are present. Calculate the mass of strontium in a 5-mCi sample of ^{90}Sr, whose half-life is 28.8 years.

10. Start with Equation 45.8 and derive the following useful formulas for the decay constant and half-life:

$$\lambda = \frac{1}{t}\ln\left(\frac{R_0}{R}\right) \qquad T_{1/2} = \frac{(\ln 2)t}{\ln(R_0/R)}$$

11. Verify the accuracy of the conversion factor used in Equation 45.14. In addition to the defined value for u, you will need the following constants: $c = 2.99793 \times 10^8$ m/s ; $e = 1.60219 \times 10^{-19}$ C.

12. Show that the inverse reactions $^{10}_{5}B(\alpha, p)^{13}_{6}C$ and $^{13}_{6}C(p, \alpha)^{10}_{5}B$ have the same absolute value of Q.

13. (a) What fraction of the space in a tank of hydrogen (H_2) gas at STP is occupied by the hydrogen molecules? Assume each hydrogen atom is a sphere of diameter 0.10 nm, and each molecule consists of two such spheres in contact. (b) What fraction of the space within a single hydrogen atom is occupied by its nucleus?

14. Copper as it occurs naturally consists of two stable isotopes, ^{63}Cu and ^{65}Cu. What is the relative abundance of the two forms? In your calculations, take the molar mass of copper to be 63.55 and take the masses of the two isotopes to be 62.95 u and 64.95 u.

Chapter 46

1. Draw a vector diagram showing **F**, **v**, and **B** for the Lorentz force acting on a positive ion spiralling toward one of the field coils in Figure P46.1. Note that the axial component of **F** reflects the ion back toward the center of the magnetic bottle.

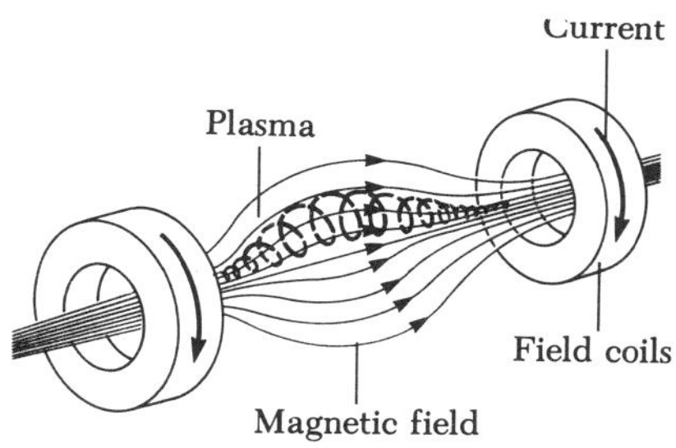

Figure P46.1

2. Inhomogeneities in the confining magnetic field make it difficult to contain plasmas. (a) Show that the magnetic field within a toroid is inversely proportional to r, where r is the distance from the central axis of the toroid. (b) Show that the fractional inhomogeneity $(B_{max} - B_{min})/B_{avg}$ in a toroidal field having rectangular windings as shown in Figure 32.32 is approximately equal to $(b - a)/r$. (c) Evaluate the fractional inhomogeneity for a toroid with $b - a = 10$ cm, first with $r = 1$ m and then with $r = 10$ m.

3. In terms of biological damage, how many rad of heavy ions is equivalent to 10 rad of x-rays?

4. (a) Estimate the volume of space required to store the radioactive wastes that would be produced in one year if all the annual U.S. electricity production (which is about 2.2×10^{12} kWh/year) came from uranium enriched to 3 percent ^{235}U. (Assume that the conversion efficiency is 30 percent and that the waste is in the form of a liquid with a density of $1\ g/cm^3$.) (b) If the waste could be formed into a cube, what would be the length of the cube's sides?

Chapter 47

1. Occasionally, high energy muons will collide with electrons and produce two neutrinos according to the reaction $\mu^{+} + e^{-} \rightarrow 2\nu$. What kind of neutrinos are these?

2. When a high-energy proton or pion traveling near the speed of light collides with a nucleus, it travels an average distance of 3×10^{-15} m before interacting. From this information, estimate the time for the strong interaction to occur.

3. (a) Show that baryon number and charge are conserved in the following reactions of a pion with a proton.

$$\pi^{+} + p^{+} \rightarrow K^{+} + \Sigma^{+} \qquad (1)$$

$$\pi^{+} + p^{+} \rightarrow \pi^{+} + \Sigma^{+} \qquad (2)$$

(b) The first reaction is observed, but the second never occurs. Explain these observations.

4. For the following two reactions, the first may occur but the second cannot. Explain.

$$K^{0} \rightarrow \pi^{+} + \pi^{-} \qquad \text{(can occur)}$$

$$\Lambda^{0} \rightarrow \pi^{+} + \pi^{-} \qquad \text{(cannot occur)}$$

5. The quark composition of the proton is uud, while that of the neutron is udd. Show that the charge, baryon number, and strangeness of these particles equal the sums of these numbers for their quark constituents.

Chapter 47

6. What is the electrical charge of the baryons with the quark compositions (a) $\overline{u}\overline{u}\overline{d}$ and (b) $\overline{u}\overline{d}\overline{d}$. What are these baryons called?

7. A gamma-ray photon strikes a stationary electron. Determine the minimum gamma-ray energy to make this reaction go:

$$\gamma + e^- \rightarrow e^- + e^- + e^+$$

8. Two protons approach each other with equal and opposite velocities. What is the minimum kinetic energy of each of the protons if they are to produce a π^+ meson at rest in the reaction

$$p^+ + p^+ \rightarrow p^+ + n + \pi^+$$

Answers to Problems

CHAPTER 1

1-1. 2.33 cm

1-2. (a) 7.14×10^{-2} gal/s (b) 2.7×10^{-4} m^3/s (c) 1.03 h

1-3. the equation is dimensionally correct

1-4 LT^{-3}

1-5 $7.46\times10^{-4}\,m^3$

1-6 4.1×10^9 y

1-7 6.71×10^8 mi

1-8 384 ft^2 $35.7 m^2$

1-9 200 km

1-10 about 10^{10} containers, or 10^5 tons

1-11 1500 times larger

1-12 (a) C = 22 cm (b) $A = 67.9$ cm^2

CHAPTER 2

2-1 (a)2.50 m/s (b) -2.27 m/s (c) $\bar{v} = 0$

2-2 -0.60 m/s

2-3 -2.5 m/s^2

2-4 1.6 m/s^2, 0.8 m/s^2

2-5 26,500 m/s^2

2-6 870 m

2-7 -4.76×10^{-5} m 15.9×10^{-9} s

2-8 234 m

2-9 (a) $-3.2\ m/s^2$ (b) 5s

2-10 23.8 s

2-11 $-32.3\ ft/s^2$

2-12 (a) 0.8 m/s (b) 18.0 m

2-13 420 m

2-14 (a) $10\ m/s^2$ (b) $x = (20 + 5t + 5t^2)$ m (c) $v = 5$ m/s

2-15 3.250 m/s, 3.125 m/s, 3.025 m/s, 3.0025 m/s, 3.00025 m/s

2-16 (a) $29.5\ m/s^2$ (b) $24.5\ m/s^2$

2-17 1.62 cm/y

2-18 $10.4\ m/s^2$

2-19 (a) 2, 4, 3, 5, 5, 2, 3, 4, 5, 2 m/s^2
(b) 0, 0.25, 1.25, 3.12, 6, 10.1, 15.1, 20.8, 27.3, 34.5, 42.6 m

2-20 4.42 m

2-21 $x(5.7\ s) = 44.3$ m, $v(5.7\ s) = 19.8$ m/s

CHAPTER 3

3-1 (5.83 m, 121°)

3-2 25 m@ ±37°, 61.8 m@ ± 14°

3-3 (a) 5.00 at 307° (b) 5.00 at 53.1°

3-4 (a) 7 blocks (b) 5 blocks, at 143°

3-5 82.5 m at 147°

3-6 (a) 6**i** +7**j** (b) 9.22 at 49.4°

3-7 (a) 2.24 (b) 8.77

3-8 −4**i** +15.4**j**

3-9 5.83 m at 149°

3-10 (−3**i** −5**j**)m, (−**i** +8**j**)m, (2**i** +13**j**)m

CHAPTER 4

4-1 (a) (5 **i** +3t **j**) m/s (b) (10m, 6m), 7.81m/s

4-2 3.19 s, 36.1m/s, −60.1°

4-3 74.5 m/s

4-4 (a) 1.03s (b) 8.67 m/s, 4.2 m/s (c) 25.8°

4-5 (a) 27.1m/s (b) 3.91s

4-6 38.3m/s, 96.4m

4-7 2.7 m/s^2

4-8 32 m/s^2

4-9 (a) $12.6 m/s$ (b) $395 m/s^2$

4-10 (a) $0.6\ /s^2$ (b) $0.8 m/s^2$ (c)$1.0 m/s^2$ at 53.1° inward

4-11 (a) $32 m/s^2$ (b) $55.4 m/s^2$ (c) $64 m/s^2$

4-12 (a) 110 km/h north (b) 110 km/h south

4-13 $0.85 m/s$

4-14 (a) 14.5°; (b) 194 km/h

4-15 18.2 m

4-16 $15.6 m/s$

4-17 $8.94 m/s$ at -63°

4-18 87.3m upstream , 567.3m downstream

4-19 He must long jump to reach safety with 3.18 ft to spare

4-20 utilize Simpson's method to find x, y, and v.

4-21 (a) $\Delta r = 0.511 m$ (b) $t = 0.74 s$

CHAPTER 5

5-1 3 m/s^2

5-2 (a) $5 m/s^2$ (b) 19.6 N (c) $10 m/s^2$

5-3 (a) 10^5 dynes (b) 0.225 lb

5-4 1.8 N

5-5 70.5 kg

5-6 6N East, 6 kg

5-7 7 N to the left

5-8 (a) 15lb (b) 5lb (c) 0

5-9 900 kg, -1.36×10^5 N

5-10 (a) $-2.5\ \mathrm{m/s^2}$ (b) 2500 N

5-11 (a) $T = 36.8$ N (b) $a = 2.45\,\mathrm{m/s^2}$ (c) $x = 1.23\,\mathrm{m}$

5-12 3.73 m

5-13 112 N

5-14 $t = 7.0$ s

5-15 0.727, 0.577

5-16 0.456

5-17 1.66×10^6 N

5-18 (a) $2.00\,\mathrm{m/s^2}$ (b) 4N, 6N, 8N (c) 14.0 N, 8.0 N

5-19 (a) $1.02\,\mathrm{m/s^2}$ (b) 2.04 N, 3.06N, 4.08N (c) 14N, 8N

5-20 Proof follows from $\sum F_y = 0$ and $\sum F_y = 0$.

CHAPTER 6

6-1 60.0 N

6-2 0.283

6-3 2.87 N down

6-4 $2.42\ \mathrm{m/s^2}$

6-5 (a) $6.27\,m/s^2$ downward (b) 784 N directed up (c) 283 N

6-6 (a) $51.7\,m/s$ (b) $137\,m$

6-7 $0.493\,m/s^2$ $-0.524\,m/s^2$ $0.720\,m/s^2$

6-8 37.5 N

6-9 (a) 0.610 rev/s (b) $0.766\,m/s$ $2.93\,m/s^2$

6-10 $20.3\,m/s$

6-11 (a) 66.3 N (b) 36.6 N (c) 6.96 N

6-12 108 N, 55.7 N

CHAPTER 7

7-1 2.93 kJ

7-2 15.0 MJ

7-3 $0.737\ ft \cdot lb$

7-4 (a) 3.61 (b) 33.8°

7-5 32 J

7-6 3.00 J

7-7 -12.0 J

7-8 1.59 kJ

7-9 (a) 0.938cm (b) 1.25 J

7-10 (a) 33.8 J (b) 135 J

7-11 2.45×10^{10} J

7-12 $10.8\,m/s$

7-13 5.18 N

7-14 30.9 m/s

7-15 3.27 kW

7-16 (a) 131 kJ (b) 1.64 kW

7-17 (a) 29.7 kW (b) 37.3 kW

7-18 (a) 758 J (b) 454 J (c) 0 J (d) −304 J

7-19 (a) 980 J (b) −980 J (c) 24.5W

7-20 1280 J

7-21 141 kW

7-22 (a) mg^2t (b) 960 W, 1.92 kW, 0

7-23 509 J

CHAPTER 8

8-1 (a) −30.0 J (b) −51.2 J (c) −42.4 J

8-2 (a) 19.6 J (b) 2.86m/s (c) 59.6 N (d) 0.816 m

8-3 (a) $(1.96y)$ J (b) 40 J (c) $(40 - 1.96y)$ J (d) 20.4 m

8-4 (a) 19.8m/s (b) 78.4 J (c) 1.00

8-5 (a) 2.29m/s (b) −15.6 J

8-6 2.00m/s, 2.79m/s, 3.19m/s

8-7 0.721m/s

8-8 (a) 0.240 J (b) 0.180 J

8-9 (a) 0.400 J (b) 0.225 J (c) 0 J

8-10 (a) $-2ax-b$ (b) $\frac{b}{2a}$

8-11 $5.05\,\text{m/s}$

8-12 (a) $\mathbf{F} = -\frac{d}{dx}(-x^3+2x^2+3x)\boldsymbol{i} = (3x^2-4x-3)\boldsymbol{i}$ (b) 1.87 and -0.535

8-13 140 kW

8-14 10.3 GJ

8-15 (a) $-2Ax+3Bx^2$ (b) $x=0$, $x=\frac{2A}{3B}$

(c) $x=0$ is a point of stable equilibrium, $x=\frac{2A}{3B}$ is a point of unstable equilibrium

CHAPTER 9

9-1 40.0 m/s

9-2 (a) $8.35\times10^{-21}\,\text{kg}\cdot\text{m/s}$ (b) $7.50\,\text{kg}\cdot\text{m/s}$ (c) $900\,\text{kg}\cdot\text{m/s}$ (d) $1.78\times10^{29}\,\text{kg}\cdot\text{m/s}$

9-3 $-0.400\,\text{m/s}$ (to the west)

9-4 $-2.67\,\text{m/s}$, west

9-5 (a) $3.33\,\text{m/s}$ (b) -83.4 J Energy lost

9-6 (b) 0.398%

9-7 12.5 kN

9-8 $-0.500\,\text{m/s}$, $2.00\,\text{m/s}$

9-9 65.2 m/s

9-10 (a) $n\approx 42$ collisions (b) $n\approx 14$, about one third as many

9-11 $v_A = 0, \quad v_B = 2v_0$

9-12 (a) $\frac{v_0}{\sqrt{2}}$ (b) $45°$ $\quad -45°$

9-13 (a) $\mathbf{v} = -(34.3\,\text{m/s})\mathbf{i} - (25.7\,\text{m/s})\mathbf{j}$ (b) 7.71 kJ

9-14 0.375m, 1.50m

9-15 $\frac{3}{2}R$

9-16 $x_c = 0.613\text{m}, \quad y_c = 0$

9-17 (a) 1.43 m/s to the right (b) 10kg · m/s to the right

9-18 (a) 8000 kg/s (b) 6.91 km/s

9-19 $R/14$ from center of sphere

9-20 (a) 0.188 m/s (b) 3.44 km

9-21 450,000 N, $1.84g$, 2750 m/s = 6150 mi/h

9-22 (b) Set $\frac{df}{dm_m} = \frac{4m_m}{(m_n + m_m)^2} - \frac{8m_n m_m}{(m_n + m_m)^3} = 0$

This is a maximum when

$(m_n + m_m) = 2m_m$, or when $m_n = m_m$.

CHAPTER 10

10-1 (a) $2at + 3bt^2$ (b) $2a + 6bt$

10-2 $-12.7\,\text{rad/s}^2$, $3.14\,\text{s}$

10-3 $25\ \text{rad/s}^2$

10-4 158 kJ

10-5 $\frac{3}{4}ML^2$, $\frac{ML^2}{2}$, $\frac{5}{4}ML^2$

10-6 substitute M, L, x into Equation 10.16

10-7 (a) $29.6\ \text{N}\cdot\text{m}$ (b) $35.6\ \text{N}\cdot\text{m}$

10-8 (a) $21.6\ \text{kg}\cdot\text{m}^2$ (b) $3.60\ \text{N}\cdot\text{m}$ (c) 52.4 rev

10-9 12.6 kW

10-10 (a) $2.00\ \text{rad/s}^2$ (b) $6.00\,\text{rad/s}$ (c) 90.0 J (d) 4.50 m

10-11 $\frac{2}{5}MR^2$

10-12 (a) $T = \frac{Mmg}{M+4m}$ (b) $a = \frac{4mg}{M+4m}$ (c) $\sqrt{\frac{8mgh}{(M+4m)R^2}}$

CHAPTER 11

11-1 $\frac{15}{14}$

11-2 $(-7)\mathbf{i} + (16)\mathbf{j} + (-10)\mathbf{k}$

11-3 $|\mathbf{F}_3| = |\mathbf{F}_1| + |\mathbf{F}_2|$

11-4 (a) $4dmv\mathbf{k}$ (b) 0 (c) $-2dmv\mathbf{k}$

11-5 (a) $2.10\times10^{10}\ \text{kg}\cdot\text{m}^2/\text{s}$ (b) L is constant in level flight

11-6 (a) $3.14\text{N}\cdot\text{m}$ (b) 0.40v (c) 7.85m/s^2

11-7 (a) $0.32\text{kg}\cdot\text{m}^2$ (b) $0.96\ \text{kg}\cdot\text{m}^2/\text{s}$ / (c) 4.80 N

11-8 (a) 0.61 rad/s (b) 0.122 J

11-9 (a) $3.50\text{kg}\cdot\text{m}^2/\text{s}$ (b) 196 J

11-10 (a) $\sum T = I\alpha$ (b) $\sum F = ma$

11-11 $60t\ \mathbf{k}\ \text{kg}\cdot\text{m}^2/\text{s}$, $-60\mathbf{k}\ \text{N}\cdot\text{m}$

11-12 $\boldsymbol{L} = \boldsymbol{r}\times\boldsymbol{p}$

11-13 (a) $v_{cm} = \sqrt{\dfrac{4g(R^3 - r^3)}{3r^2}}$ (b) 5.31×10^4 m/s (c) The energy goes into heat

CHAPTER 12

12-1 $x = (\frac{\ell}{2} + d)\left(\frac{2}{3}\frac{W_1}{W_2} - \frac{1}{3}\right)$

12-2 $y_{cm} = 15.3$ cm, $x_{cm} = 8.00$ cm

12-3 $\frac{2}{3}L$

12-4 0.25 m to the right

12-5 706 N · m

12-6 2.24 m

12-7 2.23×10^{-5}

12-8 2.8×10^{7} N/m^2

12-9 1.097 cm^3

12-10 $1.155 Mg$, $0.577 Mg$

12-11 184 N, 187 N, 213 N

12-12 12.3 kN

12-13 7.51×10^{6} N/m^2

12-14 9.85×10^{-5}

CHAPTER 13

13-1 (a) $-12\pi \sin(3\pi t + \pi)$ m/s (b) $-36\pi^2 \cos(3\pi t + \pi)$ m/s^2
(c) 12π m/s and $36\pi^2$ m/s^2 (d) 0, $36\pi^2$ m/s^2

13-2 (a) 4.33 cm (b) -5.00 cm/s (c) -17.3 cm/s^2 (d) 3.14s

13-3 (a) 31.6 rad/s (b) 1.58 m/s

13-4 3.33 cm

13-5 (a) -1.44 m/s, -3.92 m/s^2, (b) 1.57 m/s, 9.87 m/s^2

13-6 (a) 1.19 Hz (b) 0.21 s (c) 0.784 N

13-7 (a) 126 N/m (b) 0.178 m

13-8 (a) 0.153 J (b) 0.784 m/s (c) 17.5 m/s^2

13-9 0.384 J

13-10 6.35 s, 0.158 Hz

13-11 21.9 m

13-12 1.50 m

13-13 $\ell = 2R$

13-14 $[b] = \frac{M}{L}$

13-15 The coefficients are equal, $x = Ae^{-bt/2m}\cos(\omega t + \delta)$ is a solution of Equation 13.29

13-16 differentiate $x = A\cos(\omega t + \delta)$ twice and substitute into Equation 13.32

13-17 (a) $\frac{1}{2}(M + \frac{M}{3})v^2$ (b) $2\pi\sqrt{\frac{M + M/3}{k}}$

13-18 (a) 11.3 cm (b) -0.014 N (c) 0.667 s (d) ± 21.8 cm/s
(e) $E = 1.58 \times 10^{-3}$ J $k = 0.123$ N/m

CHAPTER 14

14-1 2.97×10^{-9} N

14-2 4.01×10^{-8} N, 2.31×10^{-8} N, 4.62×10^{-8} N, 30°

14-3 $\frac{d}{3}$

14-4 6×10^{24} kg

14-5 9.37×10^{6} m

14-6 5.90×10^{4} m/s

14-7 -1.67×10^{-14} J

14-8 2.49 km/s

14-9 (a) 1.88×10^{11} J (b) 103 kW

14-10 $F = \dfrac{GML\lambda_0}{d(L+d)} + GmAL$

14-11 (a) 0 (b) 2.67×10^{-10} N

14-12 0.500

14-13 7.42×10^{-9} N at 58.5°

14-14 2.67×10^{-7} m/s^2

14-15 5.98 km/s, 8.27 km/s

CHAPTER 15

15-1 20.6 m

15-2 $F = 5.5 \times 10^4$ N

15-3 2.47%

15-4 2.66×10^4 Pa

15-5 10.3 m

15-6 1.96×10^3 N

15-7 10.7%

15-8 (a) $\rho_{\text{obj}} = 5000\,\text{kg/m}^3$ (b) $\rho_{\text{fluid}} = 667\,\text{kg/m}^3$

15-9 77.7 kg

15-10 4.44 atm

15-11 4.31×10^4 N

15-12 1.67×10^2 kg/m^3

15-13 $1.4 \times 10^8 \text{ N/m}^2$

15-14 The pressure on each side of the interface is equal

15-15 5.06 kW

15-16 58.6 kPa

15-17 $t \cong 33$ km

15-18 (a) $(\rho_1 h_1 + \rho_2 h_2)/(h_1 + h_2)$ (b) $(\rho_1 h_1 + \rho_2 h_2)/\rho_w$ (c) $d' = d$

15-19 (a) $T_\ell = \left[\frac{\rho_b - \rho_w}{\rho_b}\right] M_b g$ (b) $T_u = \frac{1}{4}[\rho_c - \rho_w]\pi d^2 hg$ (c) $T_\ell' = M_b g$

(d) 1.14×10^4 N, 1.35×10^4 N, 1.96×10^4, 2.20×10^4 N

CHAPTER 16

16-1

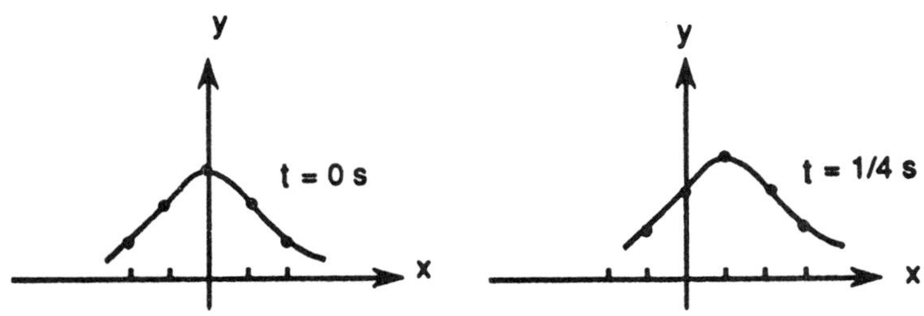

16-2 (a) − 2 cm/s

16-3 The non - physical points are $x = \pm bt/a$

16-4 8.33×10^{-2} kg

16-5 82.9 m/s

16-6 631 N

16-7 0.7 mm, 31.4 rad/m, 6.28 krad/s

16-8 (a) 0.2m (b) $4\pi\text{rad/s}$ (c) 5.03m^{-1} (d) 1.23m (e) 2.50m/s (f) left

16-9 (b) 320 cm/s

16-10 (b) 7.85m^{-1}, 2.40m/s

16-11 0.152m

16-12 (a) $y = (0.075\text{ m})\sin(4.19x - 314\,t)$ (b) P = 625 W

16-13 Both (a) and (b) are solutions

16-14 15.1 W, 3.0 J

16-15 6.67 cm

16-16 16.0m/s

CHAPTER 17

17-1 6.01 km

17-2 $[B] = \text{Pa}$

17-3 994 m/s

17-4 40 Hz

17-5 −84.3 Pa

17-6 (a) $1.92 \times 10^{-2}\text{ W/m}^2$ (b) 103 dB

17-7 3.52 dB

17-8 534 m/s

17-9 (a) 75 Hz (b) 0.948 m

17-10 18.4 m/s

17-11 5.52 μm

17-12 (a) 10 (b) 3 dB

17-13 1.79 ppm or $1.79 \times 10^{-4}\%$

17-14 34.2 W

17-15 (a) 1: 4.5: 15.5 (d) 0.331 m, 1.49 m, 5.14 m, 1.09×10^{-8} m, 1.84×10^{-10} m, 3.53×10^{-11} m, 2.92×10^{-2} N, 1.73 N, 9.00 N

CHAPTER 18

18-1 $y_2 = 8 \sin[2\pi(0.1x - 80t - \frac{1}{6})]$

18-2 (a) 0.34 m (b) 0.68 m

18-3 (a) 6.04 rad (b)1.985 A

18-4 it does

18-5 1050

18-6 120 Hz

18-7 (a) f_1 is reduced by $1/2$ (b) f_1 is reduced by $1/\sqrt{2}$ (c) f_1 is increased by $\sqrt{2}$

18-8 (a) $L = nv/2f$ (b) $L = (2n-1)\,v/4f$, $n = 1, 2, 3\ldots$

18-9 0.253 m, 0.506 m, 0.759 m,... $n(0.253)$ m

18-10 (a) 572 Hz, 1144 Hz, 1716 Hz, (b) 286 Hz, 858 Hz, 1430 Hz

18-11 (a) 215 Hz (b) 1935 Hz

18-12 (a) 21.4 m, 5.72 mm (b) 10.7 m, 2.86 mm

18-13 3400 Hz

18-14 (a) 4 Hz (b) 1%

18-15 4 Hz

18-16 (a) 133 Hz (b) 5.64 kg

CHAPTER 19

19-1 (a) 54.6 mm Hg (b) 196 mm Hg

19-2 (a) -251° C (b) 1.36 atm

19-3 40.6° C to -31.6 C

19-4 119° C, 246° F, 832° F

19-5 430 K, 774° F

19-6 1.000329 m

19-7 (a) $1.50 \times 10^{-5} (C^\circ)^{-1}$ (b) -41.7° C

19-8 2.16×10^{-3}

19-9 1.43 g

19-10 (a) 900 K (b) 1200 K

19-11 3.84 m

19-12 (a) $\beta = \dfrac{1}{T}$ (b) $3.67 \times 10^{-3^\circ} C^{-1}$

19-13 0.417 L

19-14 22.1 atm

19-15 $\bar{\beta} = \dfrac{h_t - h_0}{h_0 t_c}$

CHAPTER 20

20-1 165° C

20-2 361 pellets

20-3 (a) 0.435 cal/g·C° (b) beryllium

20-4 (a) 84.4° C (b) 557 cal gained, 998 cal gained, 1560 cal

20-5 12.9 g steam

20-6 (a) 24 g ice at 0° C (b) 8.22° C

20-7 101.3 J

20-8 (a) −12.0 MJ (b) +12.0 MJ

20-9 (a) −107 J (b) 703 J

20-10

		Q	W	ΔU
20-10	BC	−	0	−
	CA	−	−	−
	AB	+	+	+

20-11 (a) 76.0 J, 101 J, 88.6 J, (b) 167 J, 192 J, 180 J

20-12 (a) 71.2 kJ (b) 0.702 m^3

20-13 (a) (100 R) J (b) (250 R) J (c) (150 R) J

20-14 (a) 3.12 J/s (b) 9.36 mg/s (c) 1.38 mg/s (d) 500 C°/m

20-15 192×10^3 C°

20-16 (a) 3.04×10^{-2} J (b) 3.63×10^{-6} times as large

20-17 (a) $-\frac{P_0 V_0}{2}$ (b) $-1.39\ P_0 V_0$ (c) W = 0

20-18 (a) 600 kJ (b) 400 kJ (c) 1.07×10^6 J (d) 1.47×10^6 J

20-19 (a) 1.60 kW (b) 1.41 ks

20-20 285 J

CHAPTER 21

21-1 (a) 201 cal (b) 281 cal

21-2 (a) 2.99 mol (b) 8.93cal/K (c) 1.31×10^4 J

21-3 (a) 14.9 cal/K, 8.96 cal/K (b) 20.9 cal/K, 14.9 cal/K

21-4 151 kJ

21-5 3.33×10^{-26} kg

21-6 Substitute $P = \frac{nRT}{V}$ into the first equation

21-7 (a) 15.6 (b) 0.0214

21-8 10.1 atm, 759 K

21-9 A complex molecule like SO_2 contains more rotational and vibrational energy states.

21-10 (a) 6.80 m/s, (b) 7.41 m/s (c) 7.00 m/s

21-11 731 m/s, 825 m/s, 895 m/s

21-12 (a) 3.6×10^4 molecules (b) 1.8×10^4 molecules

21-13 0.15 or 15%

21-14 (a) 3.21×10^{12} molecules (b) 7.78×10^5 m (c) $6.42 \times 10^{-4}\ s^{-1}$

CHAPTER 22

22-1 (a) 1.8 kW (b) 459 J

22-2 0.305, 16.7 MJ

22-3 (a) 2.46×10^9 J (b) $+0.49$ C°

22-4 197 kJ

22-5 (a) 741 J (b) 459 J

22-6 (a) 1.75 (b) 0.80

22-7 9

22-8 -90.2 J/K

22-9 3.59 J/K

22-10 (a) -6.4 J/K (b) $+10.7$ J/K (c) 0 (d) $+4.3$ J/K

22-11 57.2 J/K

22-12 (b) Area enclosed = work done by system

22-13 (a) 53 °C or 326 K (b) 7.33 J/K

22-14 29.3 J

22-15 (a) 13.4 J/K (b) 98.19 F°, 2.59 cal/K

22-16 (a) +3.00 kJ (b) -1.00 kJ (c) +2.00 kJ
(d) +1.00 kJ (e) +1.00 kJ (f) 0.333

22-17 (a) 120 kJ (b) 3.0

CHAPTER 23

23-1 $F = 1.60 \times 10^{-9}$ N repelling

23-2 $F = 2.03 \times 10^{-3}$ N

23-3 85.0 mN

23-4 $q_1 = \pm 5.82\ \mu\text{C},\ q_2 = \mp 2.31\ \mu\text{C}$

23-5 $\dfrac{1.35(\mathbf{i}+\mathbf{j})kq^2}{a^2}$

23-6 $\tau = \sum \mathbf{r}_i \times \mathbf{F}_i = 0$

23-7 (a) $(-5200\ \mathbf{i})$ N/C (b) $2930\mathbf{j}$ N/C (c) 5850 N/C at 225°

23-8 $-54{,}000\ \mathbf{i}$ N/C

23-9 $E = 0$

23-10 $E = 0$

23-11 $E = 0$

23-12 (a) $0.914\dfrac{kq}{a^2}$ at an angle of 225° (b) $0.914\ \dfrac{kq^2}{a^2}$ at an angle of 45°

23-13 7.82 m to the left of the negative charge

23-14 (a) 0.145 C/m^3 (b) 1.94×10^{-3} C/m^2

23-15

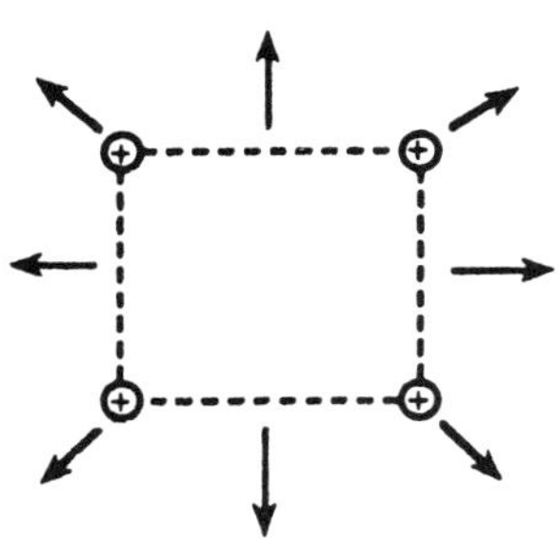

23-16 32.51 m

23-17 (a) $\theta_1 = 2\theta_2$

23-18 (a) $\frac{\pi\lambda k}{R\sqrt{2}}$ (b) $(-0.517)\frac{\pi\lambda k}{R}$

23-19 (a) 6.24×10^{-26} N · m (b) -6.24×10^{-26} J, $+6.24 \times 10^{-26}$ J

23-20 (a) $[\eta]$ = kg/m · s (b) $6\pi\eta r v_t = m_0 g - m_{air} g$

CHAPTER 24

24-1 (a) 1.98×10^6 N · m²/C (b) 0 (c) 1.92×10^6 N · m²/C

24-2 $\frac{chw^2}{2}$

24-3 $\Phi = \frac{bhw^2}{3}$

24-4 flux entering cone = flux leaving cone

24-5 5.65×10^5 N · m²/C

24-6 -4.6×10^5 C

24-7 13.7 μC, No

24-8 $Q = -5.56 \times 10^{-8}$ C

24-9 (b) 0 (c) 4.52×10^5 N · m²/C

24-10 (a) 355 N · m²/C (b) 2.84 kN · m²/C (c) 2.84 kW · m²/C

24-11 $(7.64 \times 10^4 \hat{\boldsymbol{r}})$ N/C

24-12 (a) $Q = +9.12 \times 10^{-7}$ C (b) $E = 0$

24-13 (a) 2.56×10^6 N/C (b) $E = 0$

24-14 (a) $\sigma = 1.42 \times 10^{-6}\ \mathrm{C/m^2}$ (b) 0.355 μC

24-15 The same electric field for the same charge distribution

24-16 (a) 0 (b) 12.4 kN/C (c) 639 N/C

24-17 (a) $q_{in} = 0$ (b) $q_{in} = 8.00\ \mu c$ (c) $q_{in} = 0$ (d) $q_{in} = 4.00\ \mu c$

24-18 (a) $0.496 \mu\mathrm{C/m^2}$ (b) $0.248\ \mu\mathrm{C/m^2}$

24-19 $\frac{GM_E}{R_E^3} r$

CHAPTER 25

25-1 Zero

25-2 (a) 59.0 V (b) $r = 4.56 \times 10^6$ m/s

25-3 $E = 1.56 \times 10^3$ N/C

25-4 260 V

25-5 260 V

25-6 (a) -50.0 V (b) -6.00×10^{-4} J

25-7 $V = -8.82 \times 10^4$ V

25-8 -1.83×10^5 V

25-9 0.116 J

25-10 3.53 J

25-11 20.1 J

25-12 (a) −8.00 V (b) + 8.00 V/m, −3.00 V/m, + 4.00 V/m

25-13 (a) 3.38 MV (b) 3.00 MV

25-14 (b) $1.27 \times 10^9 \mathbf{i}$ N/C

25-15 $q_1 = 4.37 \times 10^{-8}$ C

25-16 1200 W

25-17 5.12×10^4 V

25-18 (a) 0.30 m (b) 6.67 nC

25-19 7.50 W

25-20 (a) 7.20 V (b) 7.68 V (c) 7.85 V (d) 7.91 V

25-21 (a) 7.2 V (b) 6.44 V (c) 6.44 V

CHAPTER 26

26-1 $[C] = \frac{[Q]}{[V]} = \frac{C}{N \cdot m/C} = \frac{C^2}{N \cdot m}$

26-2 1.40 pF

26-3 (a) 0.819 m (b) 2.55×10^{-6} C, 3.02×10^{-7} C/m^2

26-4 0.109 m

26-5 3.54 nC

26-6 (a) 4.00 μF (b) 16.0 μC, 32.0 μC (c) 4.00 V, 6.00 V, 2.00 V

26-7 (a) 1.25 μF (b) 15.0 μC, 3.00 μC, 12.0 μC

26-8 (a) 12.0 μF (b) 24.0 μC, 14.4 μC, 19.2 μC

26-9 7.48 μF

26-10 6.71×10^{-4} J

26-11 (a) 0.15 J (b) 268 V

26-12 21.8 V/m

26-13 $U = \frac{1}{2} \varepsilon_0 E^2$

26-14 (a) 8.13 nF (b) 2.40 kV

26-15 10.1 V

26-16 26.8

26.17 3.00 μF

26.18 (a) 1.73 m^2 (b) 3.60 J

CHAPTER 27

27-1 0.481 μA

27-2 33.3 C

27-3 5.90×10^{28} e/m^3

27-4 $3.23 \times 10^6 (\Omega \cdot m)^{-1}$

27-5 26.7 Ω

27-6 2.56 mm

27-7 21.0 Ω

27-8 $R = 15.7\ \Omega$

27-9 7% decrease

27-10 1.711 Ω

27-11 $I = 500$ A

27-12 V = 82.9 V

27-13 (a) 360 kJ (b) 100 W

27-14 (a) 13.75 Ω (b) 7.39 A

27-15 (a) $2.55 \times 10^{-5}\ \Omega$ (b) 75.8 kg

CHAPTER 28

28-1 $R = 100\ \Omega$

28-2 9.52 Ω

28-3 0.545 R

28-4 The top, center, and bottom resistors are at the same potential.

28-5 (a) −5.00 V (b) −9.00 V (c) −15.0 V

28-6 (a) $I = 0.67$ A (b) Q = 50.0 μC

28-7 $i_1 = 2.67$ A, $i_2 = 2.5$ A, $i_3 = 167$ mA

28-8 (a) 6.5×10^{-3} J (b) 6.5×10^{-3} J

28-9 1940 Ω

28-10 (a) 2.48 A (b) 4.00 V

28-11 (a) 0.500 A (b) 4.63 V (c) 4.25 V (d) 450 J

CHAPTER 29

29-1 2.31×10^{-13} N

29-2 (a) 7.90×10^{-12} N (b) 0

29-3 $(12.3\mathbf{i} + 4.48\mathbf{j} - 1.60\mathbf{k}) \times 10^{-18}$ N

29-4 16.9 N/m

29-5 $\mathbf{B} = (0.042\,\text{T})\mathbf{k}$

29-6 (a) 9.00 mN, the force points 15° above North
(b) F = 0.00233 N, The force is into the page, or West

29-7 $1.84 \times 10^{-2}\ \text{A} \cdot \text{m}^2$

29-8 0.0589 J

29-9 (a) 3.02 m (b) 7.15×10^{-7} s (c) 1.4×10^{6} Hz

29-10 $B = 1.82 \times 10^{-4}$ T

29-11 (a) 3.44×10^{7} m/s (b) 20.5 ns (c) 6.20 MeV

29-12 (a) 0.614 T (b) 9.37 MHz (c) 1.23 T (d) 21.0 MeV

29-13 3.69×10^{-7} V

29-14 (a) **k** (b) 2.09 m (c) 3.28 m (d) 1.64×10^{-8} s

29-15 $\lambda\omega\,\dfrac{\ell^3}{24}$

29-16 3.13 cm

29-17 31.8 μs

CHAPTER 30

30.1 47.7 turns

30.2 (a) 189μT (b) 7.66 cm

30.3 note that $dV =$ Ads and that $J = I / A$

30.4 8.00×10^{-4} T

30.5 10.0 μT

30.6 5.00 cm

30.7 26.7 mT, 37.7 mT

30.8 (a) $B\pi R^2 \cos(\omega t)$ (b) 0

30.9 J/m^3

30.10 χ =206, ferromagnetic material

30.11 366 g/mol

CHAPTER 31

31-1 B = 0.180 T

31-2 2.67 T/s

31-3 1.32 cos (60πt) V

31-4 $\frac{2\pi}{3} R^2 N B_0 \omega \sin \omega t$

31-5 2.81×10^{-3} T

31-6 (a) $\frac{1}{2}$A (b) 2.00 W (c) 2.00 W

31-7 (a) 1.18 mV

31-8 (a) $\frac{\omega B}{8} L^2$ (b) 0.36 V

31-9 (a) 3.35 V (b) 3.72 V

31-10 2.23×10^{-5} N/C

31-11 7.07 rad/s

31-12 12.0 turns

31-13 (a) (1.19 V) cos $(120\pi t)$ (b) 88.5 mW

31-14 (b) 0.250 T

CHAPTER 32

32-1 4.69 mH

32-2 $[L] = \frac{\text{N} \cdot \text{m}}{(\text{C} / \text{s})^2}$

32-3 1.1×10^{-3} H

32-4 0

32-5 2.61 H

32-6 3.0×10^{-2} H

32-7 $[L]/[R] = \frac{V / (\text{A} / \text{s})}{V / \text{A}} = \text{s}$

32-8 (a) 2.00 ms (b) 0.176 A (c) 1.50 A (d) 3.22 ms

32-9 3.33 ms

32-10 92.8 V

32-11 (a) 16.7 W (b) 28.8 W (c) 7.19 J

32-12 (a) 0.500 J (b) 17.0 W (c) 11.0 W

32-13 $(0.112 - 0.168\ t)$V

32-14 (a) $\pi\mu_0 N_1 N_2 \frac{r_2^2}{\ell}$ (b) $\pi\mu_0 N_1 N_2 \frac{r_2^2}{\ell}$ [same]

32-15 (a) 5.19 mA (b) 1.72 MHz (c) 7.68×10^{-9} J

32-16 $t = (\frac{2n+1}{8})T;\quad n = 0,1,2,\ldots$

32-17 (a) True (b) overdamped RLC circut

32-18 $\varepsilon = -L\frac{dI}{dt}$

32-19 (a) 30 V (b) −300 A/s

32-20 (a) 56.9 W (b) 36.0 W (c) 20.9 W

CHAPTER 33

33-1 (a) 1.67 A (b) 314 rad/s

33-2 (a) −1.44 A (b) 1.45 A

33-3 $X_L = \omega L$

33-4 76.8 mH

33-5 $X_C = 1/\omega C$

33-6 335 Ω

33-7 $f = 61.6$ Hz

33-8 $C = 5.49 \times 10^{-7}$ F

33-9 1.43 W

33-10 (a) 0.627 (b) 69.0 mA (c) 4.27 W

33-11 (a) 0.591, 0.987

33-12 (a) 5.0 A (b) 2.77 A (c) 2.77 A

33-13 (a) 0.9998 (b) 0.0199

33-14 VI = constant

33-15 240 V (rms), 2.38 A

33-16 468 V

33-17 (a) 16.1 kΩ (b) 0.155 mA

33-18 302 mH

CHAPTER 34

34-1 $c = 1/\sqrt{\mu_0 \varepsilon_0}$

34-2 $B = 1.77 \times 10^{-11}$ T

34-3 $\dfrac{\partial B}{\partial x} = -\mu_0 \varepsilon_0 \dfrac{\partial E}{\partial t}$

34-4 0.201 W/m^2

34-5 (a) 8.01×10^{-7} T (b) 153 W/m^2 (c) 2.55×10^{-7} J/m^3
(d) 2.55×10^{-7} J/m^3

34-6 2.90 A

34-7 (a) $1.60 \times 10^{-10} \frac{\text{kg} \cdot \text{m}}{\text{s}}$ (b) 1.60×10^{-10} N

34-8 (a) 5.7×10^{8} N (b) 3.53×10^{22} N

34-9 (a) $P = 0.933\, P_{max}$ (b) $P = 0.500\, P_{max}$ (c) $P = 0$

34-10 5.45×10^{14} Hz

34-11 (a) 3.00×10^{10} Hz (b) 3.00×10^{14} Hz (c) 5.17×10^{14} Hz
(d) 3.00×10^{15} Hz (e) 3.00×10^{20} Hz

34-12 $S = \frac{EB}{\mu_0}$

34-13 1.2 MJ

34-14 (a) 3.51×10^{4} photons/m^3 (b) 4.99×10^{-11} J/m^3

34-15 5.16×10^{-10} T

CHAPTER 35

35-1 2.998×10^{8} m / s

35-2 60 μs

35-3 94.6°

35-4 0.890

35-5 3.4 m

35-6 1.50

35-7 7.89°

35-8 (a) 18.6° (b) 6.20°

35-9 13.2 m

35-10 77.5°

35-11 48.6°

35-12 $t = t_1 + t_2$

35-13 geometry of complementary angles

35-14 (a) 45° (b) 22.5°, 11.25°, etc.

CHAPTER 36

36-1 (c) seven images

36-2 0.20 m

36-3 (a) $q = -13.3$ cm, M = 0.667 (b) $q = -24.0$ cm, M = 0.400

36-4

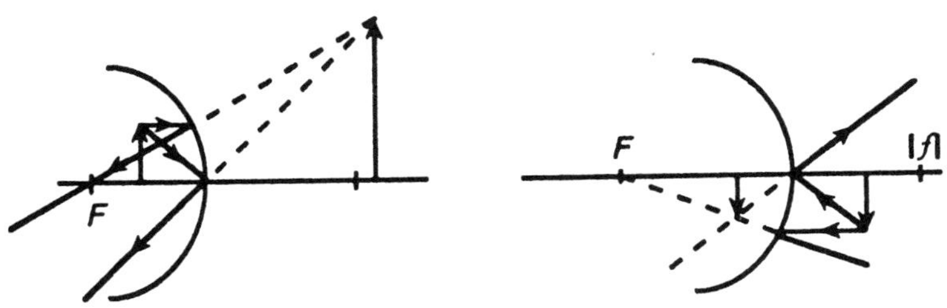

36-5 −25 cm

36-6 (a) −13.33 cm (b) −9.23 cm (c) −3.79 cm

36-7 $f = -0.484$ m

36-8 −40 cm

36-9 0.218 cm

36-10 +2.89 diopters

36-11 −0.5 diopters

36-12 (a) 24.5 cm (b) 99.0 cm

36-13 $m \to +1$ when $f \to \infty$

36-14

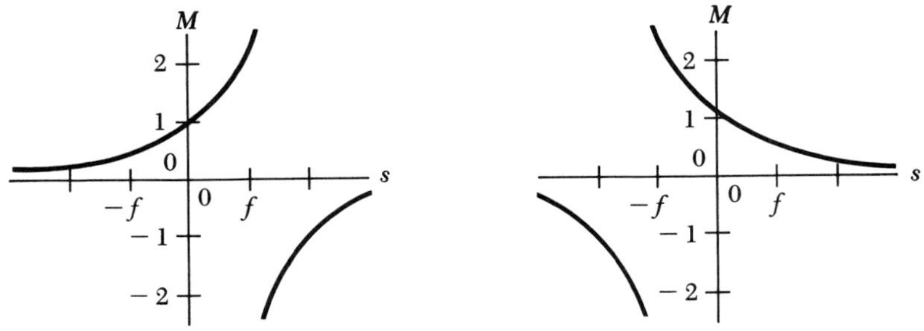

36-15 (b) −6.67 cm, −20 cm, 40 cm, 30cm

36-16 (a) $\frac{4}{3}f$ (b) $\frac{3}{4}f$ (c) −3.00, + 4.00

CHAPTER 37

37-1 0.340 m

37-2 4.80×10^{-3} m

37-3 4.48×10^{-3}%

37-4 1.56×10^{-4} m

37-5 38.1

37-6 94 nm

37-7 2 mm

37-8 0.0162 m

37-9 (a) 2A (b) 1.73 A (c) 1.41 A

37-10 (b) Green

37-11 1.7

37-12 for each 1 cm of vertical displacement, thickness varies by 492 nm.

CHAPTER 38

38-1 (a) $0.105°$ (b) 2.20 mm

38-2 $\phi = \beta / 2 = 4.4934,\ 7.7252,\ 10.904$ rad.

38-3 (a) 1.03 cm (b) 20.6 cm

38-4 2.5×10^{-7} rads

38-5 (a) $5.92°$ (b) $6.44°$

38-6 4.03×10^{-7} m

38-7 (a) 2630 slits/cm (b) 433 nm

38-8 151 μrad, 312 μrad, 499 μrad

38-9 $19.5°$

38-10 it is not possible

38-11 2.82×10^{-10} m

38-12 $\frac{3}{8} I_0$

38-13 42%

38-14 660 nm

38-15 0.2197 rad, 0.3846 rad, 10.0 cm

38-16 (b) 0.0553 deg/nm

38-17 $R = \frac{\lambda}{\Delta\lambda}$ and $\sin\theta \approx \frac{\Delta y}{L} = \frac{m\Delta\lambda}{a}$

CHAPTER 39

39-1 0.866c

39-2 $v = 0.373c = 1.12 \times 10^7$ m/s

39-3 1.54 ns

39-4 (a) 2.2×10^{-6} s (b) 649 m

39-5 0.1404 c

39-6 4.50×10^{-14}

39-7 $E^2 = p^2c^2 + (mc^2)^2$ and $K = (\gamma - 1)mc^2$

39-8 0.786 c

39-9 4.98 MeV

39-10 (a) 64.9 min^{-1} (b) 10.6 min^{-1}

39-11 (a) 0.0236c (b) 6.18×10^{-4} c

39-12 $u = 0.8$ c

39-13 (a) 8.33×10^{-8} s (b) 23.7 m

39-14 6.17 ns

39-15 5600 MeV

39-16 1.47 km

39-17 $f_0 = \dfrac{1}{T_0} = \dfrac{c - v}{\lambda_0} = \dfrac{c - v}{\lambda / \gamma} = \dfrac{(c - v)\gamma}{c / f}$

CHAPTER 40

40-1 1.85×10^{-34}

40-2 As $T \to \infty$, $e^{hc/\lambda kT} = 1 + \dfrac{hc}{\lambda kT}$

40-3 (a) 1.92 eV (b) 0.159 V

40-4 (a) 5.56×10^{14} Hz (b) 0.180 V

40-5 1.50 V

40-6 8.72×10^{12}/s

40-7 1.78 eV, 9.47×10^{-28} kg · m/s

40-8 23.4°

40-9 60.4 keV

40-10 (a) 3100 eV (b) 3.29×10^{7} m/s

40-11 288 keV

40-12 (a) 0.1203nm, 0.1212nm, 0.1224nm, 0.1236nm, 0.1245nm, 0.1248 nm
(b) 27.9eV, 103eV, 205eV, 305eV, 376eV, 402eV (c) 180°

40-13 1.88 μm, 1.28 μm, 1.09 μm

40-14 122 nm, 103 nm, 97.2 nm

40-15 $\dfrac{-122.4 \text{ eV}}{n^2}$ n = 1,2,3...

40-16 (a) 0.968 eV (b) 0.267 eV

40-17 (a) $\lambda = 4050$ nm, $\lambda = 1460$ nm (b) 0.307 eV, 0.850eV

40-18 (c) 9.11×10^{4} eV (b) 8.90 keV (c) 57.5°

40-19 $r = 3.55 \times 10^{6}$ m

CHAPTER 41

41-1 3.97×10^{-13} m

41-2 1.77×10^{-36} m

41-3 2.21×10^{-34} m

41-4 $(1.23\,\mathrm{V}^{-1/2})$ follows from $\lambda = h / p$

41-5 33.6 eV

41-6 151 V

41-7 546 eV

41-8 9.08×10^{-14} m

41-9 (a) $\Delta p = \hbar / L$ (b) $\Delta E = \hbar^2 / 8\mathrm{m}L^2$

41-10 9.6×10^{12}

41-11 $\Psi_1(x) = (2/\mathrm{L})^{1/2} \sin(\pi \mathrm{x}/\mathrm{L})$, $\Psi_2(x) = (2/\mathrm{L})^{1/2} \sin(2\pi \mathrm{x}/\mathrm{L})$, $\Psi_3(x) = (2/\mathrm{L})^{1/2} \sin(3\pi \mathrm{x}/\mathrm{L})$

41-12 (a) $\mathrm{A} = \mathrm{a}^{-1/2}$ (b) $\mathrm{B} = \mathrm{a}^{-1/2}$

41-13 $E = \dfrac{\hbar^2 k^2}{2\mathrm{m}}$

41-14 Classically, $E = \dfrac{1}{2} kA^2$ where $E = \hbar\omega\left(\dfrac{2\mathrm{n}+1}{2}\right)$

41-15 2.25

41-16 (a) 0.1955 (b) 0.3326 (c) Yes, At large quantum numbers $P \rightarrow 1/3$

41-17 (a) $\lambda = 2 \times 10^{-10}$ m, (b) $p = 3.31 \times 10^{-24}$ kg · m/s, (c) $E = 0.172$eV

41-18 (a) 0.196 (b) 1/3

CHAPTER 42

42-1 (a) $\ell = 0, \quad m_\ell = 0,$
$\ell = 1, \quad m_\ell = -1, 0, 1$
$\ell = 2, \quad m_\ell = -2, -1, 0, 1, 2$
(b) All have energy $-6.05\,\text{eV}$

42-2 -30.6 eV, -122 eV

42-3 $\int_0^\infty \psi_{1s}^2 4\pi r^2\, dr = 1$

42-4 $\ell = 4$

42-5 $\pm 2\hbar, \pm\hbar, 0$

42-6 (a) 5.05×10^{-27} J/T (b) $\frac{1}{1836}$

42-7 $[\text{Kr}]4d^{10}$, Pd

42-8 0.12 nm

42-9 0.031 nm

42-10 0.195 nm

42-11 (a) $\ell = 2, \quad s = \frac{5}{2}\hbar$ (c) $[\text{Kr}]4d^2\,5s^2$

42-12 0.389 T/m

CHAPTER 43

43-1 12.5×10^{-27} kg, 12.2×10^{-27} kg

43-2 (a) 0.95 u

43-3 (a) $E_{rot} = 0.0156$ V (b) 8×10^{-5} m

43-4 -8.39 eV

43-5 0.938

43-6 0.50

43-7 0.671 eV

43-8 (a) 4.23 eV (b) 3.27×10^{4} K

CHAPTER 44

44-1 (a) 39.01 nm (b) 41.5 nm (c) 122 nm

44-2 (a) 3.86×10^{-2} T (b) 2.9×10^{-3} T

44-3 0.028 eV

44-4 8.8×10^{6} fluxons

CHAPTER 45

45-1 0.138 nucleons/fm^3, 2.3×10^{14} g/cm^3

45-2 6.11×10^{15} N

45-3 1.440 MeV · fm

45-4 492 MeV, 8.79 MeV/nucleon

45-5 (a) 59.915428 u (b) 8.781 MeV/nucleon, very tightly bound.

45-6 (a) 0.480 MeV (b) 0.512 MeV

45-7 (a) 8.03 MeV/nucleon (b) 8.55 MeV/nucleon (c) 8.66 MeV/nucleon
(d) 7.92 MeV/nucleon

45-8 4.83×10^3

45-9 36.3 μg

45-10 (a) from $R = R_0 e^{-\lambda t}$ (b) from $T_{1/2} = \frac{\ln 2}{\lambda}$

45-11 1.49244×10^{-12} J, 931.5 MeV/u

45-12 $Q = \Delta mc^2$

45-13 (a) 2.8×10^{-5} (b) 1.4×10^{-14}

45-14 30% and 70%

CHAPTER 46

46-1

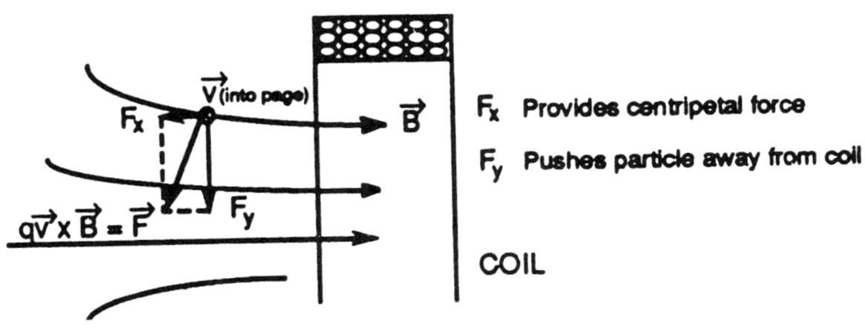

46-2 (c) 10%, 1%

46-3 0.5 rads

46-4 $V \cong 1.04 \times 10^4$ m^3, or a cube of side 21.8m

CHAPTER 47

47-1 $\bar{\nu}_\mu$ and ν_e

47-2 10^{-23} s

47-3 (b) Strangeness is not conserved

47-4 Baryon number conservation

47-5 They do.

47-6 (a) antiproton, charge -1 (b) antineutron, charge 0.

47-7 2 MeV

47-8 70.4 MeV